Starting Your Career as a CONTRACTOR

How to Build and Run a Construction Business

CLAUDIU

ALLWORTH PRESS
NEW YORK

Allworth Press books may be purchased in bulk at special discounts for sales promotion, corporate gifts, fund-raising, or educational purposes. Special editions can also be created to specifications. For details, contact the Special Sales Department, Allworth Press, 307 West 36th Street, 11th Floor, New York, NY 10018 or info@skyhorsepublishing.com.

23   7

Published by Allworth Press, an imprint of Skyhorse Publishing, Inc. 307 West 36th Street, 11th Floor, New York, NY 10018.

Allworth Press is a registered trademark of Skyhorse Publishing, Inc. a Delaware corporation.

www.allworth.com

Cover and interior design by Mary Belibasakis

Library of Congress Cataloging-in-Publication Data is available on file.

Print ISBN: 978-1-62153-458-7
Ebook ISBN: 978-1-62153-470-9

Printed in the United States of America

Table of Contents

Introduction

It is good to have an end to journey toward;
but it is the journey that matters in the end.
—Ernest Hemingway

Contractors build our homes, our offices, the roads and bridges we travel across, the parks in which our children will play, and the infrastructure we depend on for our everyday life. We build our nation one dream at a time. Our hands touch the entire world that surrounds you. The construction industry does tell many tales of arrogance, dishonesty, incompetence, and betrayal. At the core of it, however, there are many hardworking men and women who pride themselves in adhering to the highest standards. The reasons *you* seek to be a contractor are only answered by your own mind and heart, but answered they should be.

I wrote this book from the general point of view of a remodeler. Any type of contractor will find the key points mentioned here relevant to their line of work with industry-wide applications. Seasoned contractors will nod in agreement of essentials that echo in their minds. New business owners will gather useful information to help them build a solid company.

In the chapters to follow you will read about: standards and expectation on the home improvement business; how to get started on the office side and in the field; dealing with employees and clients; what to write in contracts and how to successfully bid on jobs; how to keep records and understand your profitability; how to take care of personal benefits that self-employed individuals need, and many other easily overlooked points to be careful of.

You will inevitably run into problems throughout your career. Such is life. Takeaways in this book might make some of your issues avoidable if you are open to understanding the message. Never contemplate and put yourself down. You need to solve the crisis and move on. Brighter days always wait ahead. Knowing in your heart you are always doing the right thing will make your dilemmas so much clearer. It is very important in this industry, as it is elsewhere, that you operate with integrity and honesty to yourself and others. Good communication solves most issues before they ever arise and tackles misunderstandings between you and your clients or workers. You must stay organized and focused on your business, handling paperwork promptly and efficiently. Maintain full control of your jobs, being relentless about trimming costs and improving quality. You will touch the lives of many people directly or indirectly. All of your actions will have consequences, sometimes unintended. Always strive to be better at everything you do and always seek improvement. As the U.S. Army once proudly proclaimed, *be all you can be!*

As I set out to write this book, I had imagined I would help many new contractors navigate entry hurdles in the construction business. I had also planned on helping myself evolve and systemize my thinking in business terms. Getting away from the job site gives you time to evaluate methods of streamlining your operations and make your business more efficient. I believe I have accomplished both my goals of educating others and reeducating myself. This has come at great sacrifices. As I wrote the following chapters, I was moving into a new house outside the city, we were relocating our shop and all equipment to a new location, and a fire in the building destroyed my city apartment. I had imagined a stress-free environment where I would take a few months off from working and sit on a beach writing this book. I am still dreaming. While trying to write and think clearly, I have been inundated with major remodeling jobs that my ego couldn't turn down. I unsuccessfully tried bringing my laptop on to job sites, tried hiring a manager who lost me money and turned my business into a chaotic mess, and just lost grip of my relationships with clients and employees. I do believe that everything happens for a reason that we don't yet see, and every challenge makes room for opportunity. As you go on to become your own success story, after possibly tackling many trying times, you will have gained experiences beyond the pages of this book. You will, however, find comfort in running across most of the things I describe here, giving you a leg-up on your competition.

Starting Your Career as a CONTRACTOR

Chapter 1
The Right Mindset

Why join the Navy if you can be a pirate?
—Steve Jobs

Modern society has set itself up for an unsettling scenario. We no longer live our lives. We work our lives. We spend most of our time working rather than living our lives. As much as anyone loves what he/she does for work, it still is work. Who would not trade work hours for time spent with loved ones, or time enjoying a hobby, or personal time away from the thought of work, or maybe just some extra sleep? Unfortunately, we are stuck. We spend more time with our coworkers than we do with our families. More than half of us working people wake up every morning cringing at the thought of having to go to work. Furthermore, when you have to answer to someone else, work gets even more depressing. Most of us have thought about the freedom owning your own business brings. The thought of setting your own hours and making your own decisions, while making money for yourself and not enriching someone else, is very appealing. But freedom comes with great responsibility and getting there means walking on shaky grounds. It all sounds like a dream too far from reality for most of us to achieve. The potential is, however, here for us all. I can think of no better place where entrepreneurs have achieved greatness than here in the United States, where the playing field is leveled for us all. And when it comes to construction, if you are handy and have some business sense, I can think of no other area that is more encouraging for starting your own enterprise.

Before you go off to pick a name for your newly minted business and start building skyscrapers, I would consider a few things:

- What do you really want to do in life? What makes you happy?
- Can you handle the stress of running your own business?
- Are you cut out for construction?

When the dust settles and you wake up one morning years down the road, are you excited to go to work? Planning is essential to success in life and in business. So before you jump headfirst into something new, you should consider all the alternatives and evaluate all prospects. While life is highly unpredictable, you should realize that you will encounter many challenges. This is also part of the planning. Plan for the unexpected.

You may just be starting your career and are new to the workforce. You may be changing your career by choice or because economic factors have forced you out of your current occupation. Or maybe you have been working in the construction field for some time and decided to go off on your own. Regardless of the motivation behind wanting to start your own construction business, you should know what you are getting yourself into. But most importantly, you should ask yourself if this is really what you want to do and what you wish to be doing five, ten, twenty years down the road. We grow up being molded by our environment. We look up to people we admire, while wishing to be in their shoes. Often, we follow the dreams that have been projected on to us by our parents and the society we live in. I know that you are probably excited at the prospect of starting your own construction business and you were not expecting any contrary advice in this book. Everyone's priority in this short life should be to find happiness on all levels. Success in your career comes not just from the financial freedom you may gain, but also from the work enjoyment factor that is not so easily quantified. Your overall quality of life will be altered dramatically by your career choices. With all of that being said, we are a product of our experiences. Good or bad, we should learn from life's challenges and grow on a personal level. If you ever find yourself trapped in a not-so-desirable scenario, all you have to do is move on. Life is full of choices and we are responsible for making the ones that best suit us. Make sure to do some soul searching and figure out what exactly you *wish* to be doing, and not what you *should* be doing. Before you start setting goals for your future and start working toward them, it is imperative that you understand what you want out of life. If you just go with the flow, you may just end up in a place you never intended.

Only 10 percent of newly founded construction companies survive after their first couple of years. That's not to say that 90 percent of new contractors go bankrupt. Some of them realize that running their own business is simply not for them. Many find the challenges, long hours, and stress to not pay off. The small profits incurred in the beginning of a new business make choosing a steadier career a better choice for some. Regardless of the statistics, running your own business, successful or not, is a worthy experience. It will teach you things about others and about yourself. You will learn invaluable lessons about society and the business world. Taking on such a challenge can only enhance your abilities to do even better at your next endeavor.

In 2012, Construction and Remodeling Services had 1.5 million inquiries with the Better Business Bureau, ranking at number eight for most complaints in any industry. (There are thousands of industries listed with the BBB.) If we add all the other construction specific services and businesses, we can see inquiries in the tens of millions for one year alone. These statistics not only show that there are a lot of players in our industry not equipped for the job, but also that construction is a tough market with a hard-to-please audience. In light of this fact, there is that much room for growth for the new contractor who is skilled and meticulous, and has a passion for seeing a job done right.

There is a great old saying: We are not here for a long time, but for a good time. In a nutshell, this summarizes all of the above. Always remind yourself what you are here for.

THE APPEAL

Here are many of the positive aspects of owning your own contracting business:

We all dream of being independent. After all, it is a value our country is built on. Being financially free is liberating. Running your own business has the sound of a success story and the fulfillment of The American Dream. More than half a billion new businesses are founded each year. Being your own boss is empowering, but remember that it comes with great responsibility. Working for oneself is a common ambition among contractors. Entrepreneurs who run their own construction businesses like the challenge of solving problems, and they like being in control and making their own decisions. Being in the driver's seat saves business owners from having to work for and answer to someone else. Even though you will not be going on any formal job interviews, you will still need to sell yourself and your services everyday.

By being your own boss, you create your own work–life balance. Running a business brings you the opportunity to work as little or as much as you want. You can work from and where you want to. You can live where you chose to, not where your job dictates. There is plenty of construction work available across the country. You may even have the time or the ability to incorporate some of your other hobbies into your work schedule. Some of the guys on my crew even bring their kids to work at times if the work and site are at a safe enough stage. I see no reason why you couldn't hang out with friends, family, or even your dog while you do your office work or put finishing touches on a project. Sometimes sharing your achievements with others brings even more personal fulfillment and pride. And let us not forget the fun-filled days of playing with the toughest and biggest power tools around. Big toys for big kids keep us young forever.

You make your own choices about the direction you want for your business and for your life. You also generally get to choose the people you want to work for and the folks you want to work with. Even though it may take quite a few trials, you ultimately can create your own dream team, and take on projects that stimulate your creativity and simplify your life. You can surround yourself with people who move you forward and fire those who set you back.

Owning your own construction business gives you a platform to meet many new people from a variety of backgrounds. If you are a social person, you will make a ton of new friends. Even if you are not a people person, you will be able to make some lasting connections and at least have acquaintances in many fields. You will be working with and involving many people in your community. Inevitably, you will eventually be a connecting link to individuals and businesses, from the businesses you buy supplies from to the subcontractors you employ, to the workers you hire and the clients you service. This also makes you a valuable member of your community. You can be as involved as you would like and eventually you will have the opportunity to give something back to society.

Business ownership can be very profitable on a material level and on a personal level. There is very good money to be made from renovations if you apply sound business rules and have good judgment. A home or a commercial space is built once and renovated many, many times in its lifetime. There is much more business in remodeling than there is in new construction. You can charge a very nice premium for quality work and your risk is much lower. It is also extremely rewarding to be the business owner. You will get the satisfaction of solving many problems. You will see progress in the work you have directed others to do, and you will see your projects come to completion. There is so much personal pride when you walk into a place you renovated. Years down the road you see the beautiful job that made someone happy for all this time, and you can say, "I did that!"

Also, the remodeling business is much easier to get into than new construction. The amount of time and financial investment required for building something new is extremely high. In renovations, all you need are a few tools and good work ethic. The job is sold before you even start, and generally you take a deposit up front to cover your initial expenses. If something goes wrong, you won't blow up your entire business model and go bankrupt. You just move on to the next job and the next opportunity to make money.

As with two of my previous careers, I can say that in construction there is never a boring day. (I used to be in the military and also used to work on a trading floor for a financial firm. I see many similarities between going to war, being on a trading floor, and running a contracting business.) Construction work is always challenging and ever changing. No two projects will ever be the same. This can be exciting but also very stressful for the faint-of-heart. If you are the adventurous type, then you will find yourself right at home. At the end of the day, we just have to laugh at all the crises that arise on a daily basis.

To reiterate the point I started the book with, owning your own business lets you pursue your passion. You will spend most of your time at work, so why not make it exciting. Work should be fun for you, and you should enjoy yourself. As a contractor, you should love what you do and take pride in your work. If successful, you will have done it all on your own. You started from scratch, and through your own abilities and leadership, you were able to make your business shine. There is no greater satisfaction than knowing that you are living the life you've always wanted, and through your own efforts *you* have made it all happen.

ROADBLOCKS

For all the positives you read above, there are negative aspects that come with being your own boss as a contractor. Here are a few to watch out for:

Being your own boss comes with many risks. If having a set schedule at a steady job that earns you an income you can anticipate is of importance to you, then you should probably not venture off into starting your own business. Working for yourself means that you make your own money and are responsible for finding and securing your own jobs. There is no one there to hold your hand and tell you how you need to go about your work. There is no one to hold you accountable for missing the mark. Other than, of course, your clients, you must hold yourself accountable for missing deadlines, missing cost estimates, and just missing out on work. On the bright side, if you look around in today's job market, there is no such thing as job security anywhere anymore. The quite frequent economic downturns have left many of us without jobs over and over. So maybe next time unemployment spikes over 10 percent and all your friends are out looking for a career change, you might just be able to hand them a hammer and put them to work.

In the beginning of your new construction business you will face long hours and little pay. You will have to make many personal sacrifices. Most of your relationships will be tested, whether friends or loved ones. You will find yourself up early before your crew starts work, taking deliveries or picking up materials for the day ahead. You will be the last to leave the job site as you set everything up for the next day. You will be running to other potential jobs for estimates after the workday ends. You will be at home late in the evening preparing tools and making plans for the next day. And you will spend your weekends writing proposals, sending out bids, doing payroll and taxes, and filing a lot of other paperwork. Your business will be your life. However, it is yours and only yours and there is no one else who will profit from your efforts but you.

I mentioned earlier that remodeling work can be quite profitable. I also have to remind you that it can also be quite unprofitable if the business is not run properly. If you are disorganized and you let your cost get out of control, you will quickly be out of business. Low bid estimates are also something to watch for. Getting your foot in the door *is* important, but making a living is more important. So please always consider if a job is worth taking on for the budget that's available. In the finance world, traders used to say that sometimes your best trade is the trade you never put on. So not having a job is sometimes better than doing a job that costs you money.

You will undoubtedly have to deal with difficult people. You are getting yourself into a service industry and pleasing your clients is your number one priority. Ultimately, they are the ones writing you a check. As you probably already know, there are many types of people with a wide range of personalities. You will encounter your fair share of characters, to say the least. Sometimes you will be able to spot problem clients ahead of time, and you will save yourself the hassle by not taking on that job. Other times, fires get fueled long after you've started and you must mediate situations before you get paid. More people problems will arise even from within your company. You will have to deal with employees who do not show up for work when you need them, are late for work at a critical time, or get confrontational at the wrong moment. And be ready for the subcontractor who cannot get along with your client or your crew and the work goes nowhere. In this field, interpersonal skills are important, and you will get some firsthand experiences in dealing with people.

On the same topic, there will be times when some of the work will get messed up. Something needs to be done a certain way, and when you check the job at the end of the day, it is not up to standard. Your own workers and subcontractors make mistakes or just forget what they were told. It is your responsibility to fix the problem or redo the work. Unfortunately it is also at your cost. This will happen and you should plan for it.

I say the following with deep regret. The government makes it very tough for small businesses. I love our country and believe this place still is the land of opportunities. However, our government makes it really tough for us small entrepreneurs to make good money in the beginning. As you will learn soon, there are way too many taxes and filing requirements by federal, state, and local governments.

There are also strict rules about insurance, bonds, worker's comp, and disability benefits that business owners must adhere by. All of these cost us thousands of dollars up front, before we even get to see a paycheck. And after we do get some cash flow, somehow most of it evaporates into thin air at the end of each month and each quarter. You will learn to prioritize things and set money aside for required business filings. But just as a warning, be prepared to be shocked on how much of your hard earned money has to go away to entities that do not seem to be supporting your business but just profiting from it. On top of money flushed down the drain, the amount of paperwork required of small businesses is overwhelming. From payroll filings to sales tax filings, to federal and state tax filings, to unemployment insurance, to loads of other seemingly worthless bureaucratic waste of trees, you are in for a treat. Of course, you have the option to hire an accountant, but you wouldn't quite understand your business entirely. I prepare all of my own paperwork and have been quite successful at it. In this modern age of the Internet where all the information is readily available for research, if someone else could learn to do something, then why couldn't I?

To link the above together, you should be aware that in the initial stages of your new business, you will need to do everything or at least understand all aspects of your business. If you do not have a solid understanding of the pieces that comprise your construction company, then you are not in control. The empire collapses under a ruler who loses control. You will of course have to understand how to frame a wall and how to tile a backsplash. Basic construction skills are a must, and even if you do not perform most of the daily duties, you should understand how they are done. When watching your guys install a sink, or a subcontractor pour concrete to level a floor, you have to know if what they are doing is right. If you do not, then you may get a call back for a leaking sink and a cracked subfloor. Generally, especially in the beginning, the contractors work hand in hand with their crews to see the job done. You will be a day laborer during the workday, a sales person at lunch, a financier and lawyer in the afternoon, and an accountant on the weekends. You must be comfortable wearing all these hats or have trust in others to have your best interest. You still own all the parts of your business so you must still understand how they all fit together. I personally see no other person out there that has the best interest of my company but me. So I choose to have a closet full of hats for different occasions.

I must not omit one very important drawback of owning your own business: The job comes with no benefits; you must make your own. No one will offer you and your family medical benefits. You will not be given paid vacations. There is no such thing as a retirement plan. All of the above must be thoroughly planned by you. You do, however, have some options, and you should always plan ahead. Your income will also vary widely depending on how busy you are and the type of jobs you take on. So you must balance your fruitful business times with the sedentary workdays.

I do not live in fear. I have a strong belief in making my own destiny and have full faith in my abilities to weather any storm. Survival is a choice. I would advise against letting fear guide your decisions. Maybe you are too busy to start a business. Maybe you are in debt. Maybe you are scared of failure. Maybe you know others with failing businesses. I, myself, have had several other business ventures that have failed. So what? I would much rather live with the experience of having failed at something than living in regret of not ever trying it. As Michael Jordan wisely put it: "I've failed over and over and over again in my life and that is why I succeed." There is nothing out there that I built myself that I cannot build again. I try, I fail, I start over and build it again, but only better. To myself I say: take the challenge, fail, learn from it, grow a stronger person, and live life to the fullest. There are many roadblocks in business, as in life. The way we handle them as they come is what differentiates us.

IMPORTANT IN THIS BUSINESS

You are excited at the prospect of running your own contracting business. You should be. You also know many of the challenges that must be tackled in order to be successful. As you start to plan your new venture, you should be aware of some essential skills. Some are standards that you should set for yourself and some are values you may already have and should work on improving.

First and foremost, you will need to have or develop some construction skills. You are, after all, trying to run a construction company. Your business is construction, therefore you should not only understand it, but be able to perform most tasks required on your jobs. Alternatively, you could be a fine general contractor who essentially manages projects and pulls together teams of subcontractors to perform the work. In this case you may not ever need to use a hammer yourself, but in the beginning you may not have much credibility without a track record. Most contractors start their businesses by using the skills they currently master. If you have some practice doing tile work, maybe you should look at renovating bathrooms. You may have to hire a plumber and the rest you should be able to swing on your own with a helper. This would be a starting point before you gain expertise in other areas and expand your business. Regardless, construction skills are very important in this business.

As you will be a manager, you need to understand that role and walk in those shoes. Time and money management cannot be stressed enough. You will be juggling between supervising your workers and subcontractors, organizing effective timing of hours worked, efficiently running the completion of project steps, and controlling cost. You need good overall management skills.

You are in the service industry. You will be dealing with people on a daily basis. From suppliers, to workers, to clients, your business depends on your interaction with others. Unfortunately, problems are inevitable in this field. The way in which you handle them is also critical to your success. If you are not a people's person, then you need to start getting more comfortable in social settings.

The phrase "attention to detail" always brings me back to Basic Training. Drill sergeants used to hammer that in our heads. If we were too slow to learn, they found a way we wouldn't forget. I will not be coming to your job site to make you do push-ups if you forget, but you will be reminded by your clients. Not everyone is as obsessive as I am, but the details do matter. You sometimes think that no one will notice, but they do. You might have to redo a project over or fix mistakes that were avoidable. Save yourself the embarrassment and pay attention to details.

There is no substitute for hard work. There are those who get lucky by winning the lottery, and there are those who earn it. Get-rich-quick schemes are a sure way to be out of business quickly. Those with strong work ethic will never be without a job.

Having discipline and high moral standards will also pay you dividends over time. Doing what is right and not what is easy will earn you an impeccable reputation. Contractors are notorious for sometimes being misleading. Please do not perpetuate that status. High standards will get you better work.

I cannot over-stress the importance of cost control. You must stay organized with what you spend on different projects. You should also keep them all separated so you can understand where you make money, where you bleed, and why. Controlling your expenses is absolutely vital to your business. You must always have a view of the bigger picture. Do not run into a scenario where you closed up a wall in the bathroom and tiled it only to realize you need to run an electrical wire for an outlet above the sink. Step back and see the job as a whole. Make clear to everyone the steps that need to take place, and then make sure they get executed as such. Your guys may work fast and start closing up the newly built bedroom wall. When the cable guy comes the next day to run the TV wires, all the walls have been sheetrocked. Surprise!

Time management is just as important. We have all heard horror stories about contractors who never adhere to their deadline. If you fall behind, you should always set expectations and communicate with your client. Surprises are never a good thing in construction and it is your responsibility to keep your clients happy. You have invaded the privacy of their home and have created a living hell for them. The least you can do is keep your clients informed at all times and calm their nerves. Communication and accessibility is key.

At the end of the day, you need to trust in yourself before others can put their trust in you. You should do work with a high degree of confidence and you should take pride in what you do. Well-crafted work speaks for itself and will sell your next job. If you are meticulous about everything you and your crew do, you will come out ahead. I could go on with stories of carelessly executed jobs. We often find unleveled surfaces, bumpy plasterwork, improper use of building materials, spotty plumbing, and downright code-violating electrical work. Frequently during the demolition stage when we find "treasures" hidden in the walls, my guys always smile and say, "Boss, more work for us!" This is true; if you can keep a tight grip on your operation, hire the right crew that will execute your vision, and always enforce high standards, there is plenty of work for you out there.

SOME ADVICE

As I am writing this, I realize that some of my thoughts I need not only make clear in the book, but I need to reiterate in my own mind. Some of the things I share here are actually helping me stay focused on what's important in this business. As you read some of this advice, know that I am also giving it to myself. Sometimes it helps to be reminded of what's important.

Rome wasn't built overnight. As I look around at the things I realized in my short but adventure-filled life, I realize that nothing significant ever came effortlessly. Be it relationships, business, personal accomplishments, school, or financial goals, I had to solidly work for it all. Probably nothing that comes easily is worthwhile. And the journey has generally been much more fun than the destination itself. You, as well, should expect to put in your time before you build something great. I doubt anyone thinks they can become a doctor after they took one anatomy course. So you should anticipate a long bumpy road before you get to sit on the emperor's throne and pat yourself on the back.

Be positive and upbeat. Life's stormy challenges are always easier to deal with when you see sunshine ahead. With the right attitude you can be prepared for any obstacle. Other people's problems should not be yours, so let no one bring you down.

Be a honeybee, not a drone. Your mother might have told you this at a young age, but I'll remind you right now: Don't be lazy! Sitting around your crew, eating ice cream, as they sweat in a cloud of plaster dust is not the image I want from my boss or the contractor I just hired. You have no excuse for being indolent. Challenge yourself and take chances. Life is too short. Relax on vacation, not at work. You won't get far in construction while being idle (or elsewhere in life for that matter).

Be professional. Act like you are a business owner and take responsibility when you need to. Show up on time, be up front, and keep your word. Treat others with the same respect you deserve. All of these seemingly obvious principles get overlooked much too often on too many job sites. Be the model contractor that everyone wants to work with and competitors envy!

Lastly, leave work behind. Leave your phone, your tools, and all construction-related nonsense after a certain time. Work hard during the day and go do something else afterward. If you take your mind off of work after 6 p.m., it will all still be there in the morning. Sure, there is a time to plan for the next day, or a time when you might have to pull an all-nighter. But do not make your job your life. After work you should go exercise, go to the movies, go shopping (not for construction materials or tools), go on a date, go talk to someone about the weather, or go look at the stars while listening to the crickets. Whatever you do that is unrelated to construction will help keep you sane and not burn you out. When I hear a song on the radio way too many times, I start losing interest regardless of how much I love it. My business is not a song I ever want to get sick of.

Chapter 2
Setting Up Your New Business

*By working faithfully eight hours a day
you may eventually get to be boss
and work twelve hours a day.*
—Robert Frost

While historically most of us thought of careers as traditional nine-to-five jobs that give us stability, today's world economies have changed our thinking. With high unemployment rates and sluggish growth, the job environment has been very difficult. People are also working later in life than they used to, in order to account for diminishing pensions and benefits they once had. This puts new entrants to the workforce at a severe disadvantage. In today's competitive job market, more of us are looking at opening our own business. Aside from evaluating benefits and drawbacks of owning a business, being an entrepreneur may just be the only option for some. But as you look at alternatives from the typical career, you should appropriately set your own expectations. Be prepared to work much harder for yourself and take on greater risk. Your new boss (this would be you) will be much less forgiving. As Thomas Edison said, "Genius is 1 percent inspiration and 99 percent perspiration." May you find the genius in yourself!

BUSINESS STRUCTURE

You've decided it is time to go into business with yourself. Congratulations on joining us, the 15 million self-employed Americans! Before you write CEO on your business cards, let's make certain you appropriately set up your company to prevent headaches down the road. A sound business structure will weather storms when (not if) they come. I want to be very clear that I have done all the legal and accounting steps outlined below on my own; however, I am not endorsing that anyone should do it alone. Everyone's situation is unique, and each individual should consult their own professional before making such decisions. If you have the time and willingness to study the law on your own, as I did, you may not need representation. The choice is yours, as this book is not a legal or accounting manual.

A new business owner has several options; however, for a new construction business you should really only consider the following two structures: LLC or S Corp.

Let's first run over the other alternatives and why they are not such good options.

Sole proprietor

This is the simplest form of business that has no separate structure from its owner. You are the owner and the business. You have unlimited personal liability. You will be responsible for all the debts and misfortunes of the business. You are also responsible for the actions of those working for you. Yes, go ahead and read that twice. This is called vicarious liability, meaning you can be held responsible for someone else's actions. You file Schedule C on your federal 1040 tax form and all the profits or loss of your business get reported directly on your personal tax return. You may have a very hard time getting a loan from a bank. You are liable for all mishandlings of your company or those working for it. When someone sues your company, they are really suing you personally and can go after all your personal assets, even those unrelated to your business, such as your home or your old stamp collection. Your health insurance is not tax deductible. You also may come off as an amateur handyman when you're billing someone with your personal name rather than a business identity.

The positives of being a sole proprietor are being able to file only one tax return, easy setup of the business and little ongoing formalities, quick access to your money, and the ability to freely mix personal and business assets. You are also exempt from paying unemployment tax for yourself, but of course, you get no unemployment benefits should your business suffer.

Partnership
Before I go on to say why this business structure is inappropriate in the construction industry, I want to

address those who are looking at partnering with someone. Consider still the LLC or the S Corp. Two or more partners can be members or own shares in the company. Please also reflect deeply on who you are partnering with and why. I have seen splendid relationships turn bitter. You're embarking on the journey of your livelihood, after all. Choose carefully and have formal agreements beforehand. Look at it as a marriage proposal rather than asking someone on a date.

A partnership does not protect the individual liability of the owners. The partners are equally responsible for debts and commitments of the company, and they are responsible for the other partners' actions even when not aware of them. Profits and losses are distributed by the percentage each partner owns in the company, though the liability is not shared equally. If the business goes under and your partners are broke, you may be responsible for all the debts of the company regardless of your ownership percentage. Each partner can act as an agent for the business and can borrow money, hire and fire staff, sign contracts, and make important business decisions in the name of the company (or your name for that matter). In light of these remarks, I would strongly discourage anyone from legally forming a partnership. Look at forming an LLC or an S Corp with your partners, and seek strong professional legal representation. I view a business partnership as strongly as a marriage. Be very careful before you legally bind yourself to loving someone else's actions forever.

C corporation

This is the classic corporate structure that all major businesses and publicly traded companies have. The corporation is formed under the laws of a state, which differ widely from state to state. The corporation is taxed separately from its owners, and it is required to issue financial statements. Income from a C Corp is taxed *twice*. The corporation pays tax on its net income, and then shareholders pay individual tax on distributions. The minimum federal corporate tax rate is 15 percent and can be as high as 39 percent depending on taxable income. After paying that tax at the corporate level, you then have to pay your own personal taxes for federal, state, and local governments again. In essence, you would be paying Uncle Sam considerably more than you would be paying yourself. Sounds like a lovely scenario to be in; for the uncle that is. Corporations also have additional filing requirements with the three governing bodies (federal, state, and local). Corporate law is also evolving on an ongoing basis, and such a business would inevitably require some legal and accounting advice over time. If the corporation has a loss, you may not use that against your personal income, it may only stay within the corporation. The loss can be carried forward to the next year for the corporation, but cannot be deducted from the owner's other income.

A good thing about a C Corporation is that you may leave earnings in the corporation. You do not have to take the money out as dividends. This can be helpful for purchasing equipment and property, and investing in growing your business. In this situation, your tax bill would be much lower since you'd only be paying the corporate tax on the money that never gets distributed as dividends. A soundly operated C Corp offers the most limited liability to its owners. From a legal standpoint, it should be the hardest for a court to pierce the corporate veil. [1]**

Now that these other options have been presented let's look at the two business structures you should actually consider for a construction business and why: LLC and S Corp.

Limited liability company (LLC)

This business entity is somewhat of a hybrid between a partnership and a corporation, offering liability protection as well as tax advantages and operating flexibility. The LLC can be taxed as a pass-thru entity. It passes the income through to its owners, known as members of the LLC, so they pay taxes only once at the personal level. Since the LLC is a separate entity, the members are not liable for the debts or misfortunes of the business. Some states, like CA and NY, have unfavorable tax treatments for LLCs and pay higher taxes than a corporation with similar revenue.

Small corporation (S Corp)

When owning an S Corp, you are an employee of your business and must pay yourself a salary. So you will be

---

[1] ** Piercing the corporate veil is a legal decision to treat the rights of a corporation as the rights of its shareholders. A corporation is treated as a separate legal entity, which is solely responsible for its own debts and the sole beneficiary of credits it is owed. There are instances when owners can be held liable for the corporation, negating the corporate limited liability protection.

working for your corporation, and not for yourself. At the end of the year, hopefully, you will have a business profit on top of the wage you paid yourself. The profit is distributed to you as a dividend and you pay just income tax on it. When incorporating, all businesses start as a C Corp, and then file an election to convert to an S Corp. The corporation itself does not pay any taxes, but instead the income gets distributed to its stockholders who pay individual tax on their personal returns.

Other considerations between LLC and S Corp

With an LLC, you may not deduct your health insurance, or other fringe benefits. These have to be paid after tax, out of your own pocket. With an S Corp, however, these would just be cost of doing business, all deducted from corporate income.

You have to pay self-employment tax for all your income under an LLC. With an S Corp, you only pay employment tax on your salary, while the business profit only gets taxed with income tax. You do have to pay yourself a prevailing salary, otherwise the IRS can reclassify your profits as wages and charge you the employment tax. By having an S Corp rather than an LLC, you would be saving over 15 percent in taxes on your business profit.

Different from an LLC, an S Corp does require more ongoing paperwork. As a separate structure, S Corps require scheduled director and shareholder meetings, minutes from those meetings, adoptions and updates to by-laws, stock transfers, and maintenance of records.

There is a way to combine the benefits of the two business structures by requesting S Corp status for your LLC. The LLC remains a limited liability company from a legal standpoint but for tax purposes it is treated as an S Corp. You should contact your state and a local attorney for specific tax consequences and rules.

Do not let this idea of limited liability disillusion you into believing you can be reckless and never suffer any consequences. Most likely, your business will never get a loan without personally guaranteeing for it, so you'll still be on the hook for your debt. Also, if you get into legal problems, a smart attorney will find a way to pierce the corporate veil by discovering a way you have misused your corporate accounts. The takeaway here is that you shouldn't fear the lawyers, accountants, and bankers, but you should run your business to always keep them on your good side.

Here is a chart outlining the major differences in areas you want to consider when forming the legal foundation of your new construction business:

	S Corp	LLC	C Corp	Partner-ship	Sole Proprietor
Limited liability	YES	YES	YES	NO	NO
Double taxation	NO	NO	YES	NO	NO
Self-employment taxation	NO	YES	NO	YES	YES
Benefits are deductible —health insurance, fringe benefits, etc.	YES	NO	YES	NO	NO

Who is legally responsible for your employees and your actions?	CORP	LLC	CORP	YOU	YOU

ACCOUNTING

You have now figured out the appropriate business entity for your business. If you haven't, read the above again, then go consult a lawyer and an accountant. Let us go on with setting up your venture.

Since you now are the legal owner of some sort of a business entity, it is time to give it its own foundation to stand on. You personally have checking and savings accounts, a few credit cards, health insurance (or not, but more on that later), and own other assets. Your business should also have all the above. They should be separate from your accounts, since the business is a separate entity. These accounts should always be kept separated and used only for their own purpose for simple accounting reasons and for more intricate legal reasons.

Here are five simple steps to follow after you have figured out your business's legal entity:

First: Obtain a Tax ID. You will need an Employer Identification Number (EIN) for your new business. You will need the EIN to file taxes for your business, hire employees, and establish credit and other banking. You can easily obtain it on the IRS website in minutes, by filling out some simple forms. You will also need to register your business with your state. All of this can be done online through your state's website. If you are incorporating, that can also be done online. You'll find tons of websites ready to take your money for filing articles of incorporation with your state. There is no need to pay anyone to do this for you. The forms are simple and do not take much time to fill out. Just do it yourself. You'll feel like a lawyer and have an extra few hundred dollars left in your pocket. You should also check with your state to see if your business name is already in use. Register your name with the state or county.

Second: Open a bank account. With your new EIN, you will be able to start banking under your business name. You should open a checking and savings account as well as a credit card for your business. Shop around with several banks before you commit. There should be no reason why you should pay monthly maintenance fees for banking. Constantly watch all these little charges that suck blood from your business. Your bank uses the money you deposit to lend to others and charge interest. They also charge a fee to every merchant when you use your credit card. So why would you pay them any more money out of your pocket? Don't! With so many options out there, you should get free checking, savings, and credit. A line of credit is also handy when you need cash for a bigger job or a government job that doesn't pay a deposit. You may need to sit down with a local banker to discuss credit options. For the most part, especially in the beginning of your career when you are taking on small jobs one at a time, your credit card will give you enough time (thirty days) to pay for your purchases. The deposit you collect before the jobs start should take care of your labor costs until the next check arrives. Remember to always keep your personal and business accounts separate.

Third: Register as a sales tax vendor. If required in your state and/or locality, you must collect sales tax from your customers. You must report this tax and pay it to the state periodically. Make certain this money gets transferred to your business savings account as soon as you get it, and it is available at sales tax filing time. Registering is easily done through your state's finance website. If you have any questions about how much sales tax to charge or where to send the money or even whether your services are subject to sales tax, just pick up the phone and call your state. I have called various government agencies on several occasions. Believe me, they are more than happy to take your money. If you can stand being on hold for extended periods of time, they are generally helpful in answering your questions. Remember that all the publications are available online and with a little research you can educate yourself as much as you need. Keep in mind that sales tax is not part of your income. It is just a tax that you must collect and send to the government. Be mindful when you commit prices to your clients. The total amount they have to pay will be higher by the sales tax percentage. If you forget to charge this tax, you'll end up paying it out of your profits. If you live in New Hampshire, or another tax-free state, you can skip the sales tax section.

If you live in N.Y.C., or another tax-loving place, you'll be collecting sales tax for the city and yet more sales tax for the state. If you love where you live, then pay the tax. If you don't, then it might be time to consider moving. One more consideration: if you work in multiple states, you must collect and distribute the appropriate tax to the right state.

Fourth: Set up payroll. Whether you are the only employee or you have others working for you, your business needs to declare the income, withhold employee taxes and add employer taxes, and send those taxes to the appropriate government agencies.

Running payroll gets a little complicated and forgetting certain forms or missing deadliness will cost you. I suggest hiring a professional for this task. You don't have to spend the money on an accounting firm to get personalized service. All big banks have payroll services integrated right in their websites for business accounts. If you meet certain criteria, these services are free. Even if you do not meet the minimum balance requirement or transaction flow, spending around $30 a month for easy access to all payroll will certainly pay off. The website will remind you of the coming deadlines, will know exactly what forms to send and where, and will withdraw the money from your account and send it to the government. All you have to do is input the income and taxes get calculated automatically.

You will be sending federal taxes monthly and depending on your situation and state requirements, state taxes either monthly or quarterly. Again, please remember to have the necessary funds available when they are due. Don't get trapped into draining your accounts on a job and have no money left for taxes. Balance your daily business needs with your long-term projections.

If you don't pay yourself a salary, you must send in quarterly estimates to both the federal and state governments. You can register through EFTPS (Electronic Federal Tax Payment System) online and through your state's tax and finance website. I will go into more depth on taxes and payroll procedures in the bookkeeping chapter.

Fifth: Get licensed. Your local codes dictate what you can and cannot do. Being familiar with the laws governing your jurisdiction is very important. Check with your local government on what type of licenses you must possess in order to do your intended work. Not only is being a licensed contractor important from a legal standpoint, but it will be much easier for a customer to trust you in their home when they see your license number printed on your business card. You own a construction business, not a handyman operation. You must have the proper certifications to stay in business. Aside from the required contractor's license, it is always a good idea to go further and get industry specific certifications. You'll not only look more professional and trustworthy on paper, but you'll also expand your knowledge base. I am a big believer in always learning and progressing. It keeps one's mind active. Some places require contractors to have a lead certification course before they can paint. Some require the OSHA ten- or thirty-hour training. All of these certifications will make your business more reputable and make you a better manager. Industry specific organizations exist for any type of business. Whether you have a painting, electrical, plumbing, carpentry, or general construction operation, there are national and local organizations that will offer certifications in your trade. These will give you higher credentials and open your business up to a greater network of contractors and customers. For example, NARI (National Association of the Remodeling Industry) offers many options to further your business, but they may or may not fit your needs. For many small local home renovators the cost and time involved in getting such certifications may not bring them enough extra work or they may not want to expand at all. Get your required license right away, and then explore as much contractor education as you possibly can. Think about your future and plan accordingly.

INSURANCE

In New York City, the minimum insurance requirements for a contractor to do any type of work are: contractor's license, liability insurance, worker's comp, and disability insurance. These are the four things we get asked for on every single project before we are ever allowed into any building. They are also a must for obtaining a work permit from the local department of buildings that oversees construction work in the city. You will not only need the permit to legally do the work, but it is required for the inspections during and after the work is completed. On most larger scale projects, you will be working with an architect who will acquire the proper permits. Check with your local government to find your specific requirements.

Keep in mind, once you walk into anyone's home, not only do they expect you to be insured, you should be insured anyway for your own sake. Insurance is a necessary evil. The construction industry is notorious for accidents and claims. Think about all the things that could go wrong on a job. You will, of course, do your best to minimize any risks, but Murphy's Law sometimes finds its way into your business. From your own tools and the homeowner's belongings, to the safety of your crew and that of the homeowners and their pets, to neighbor's property and the condo building or the city property, you could get a headache just thinking of the unthinkable. But the unthinkable does happen, so you should be prepared.

Insurance is not cheap. Many companies shy away from insurance for construction. You will have to shop around and do your homework. Pay close attention to the exclusions on the policy you are given. There may be restrictions on work done outside or above a certain height. There are so many different clauses that the insurance companies put in to protect themselves. You should keep in mind what and where your work is, and make sure you are covered for it. I have even seen business owners carry liability insurance that prohibits any work in New York City, while this is their primary work area. Silly, but it could be an unfortunate experience should anything go wrong. I suggest you find a few insurance brokers and discuss your needs with them. Local brokers that insure other similar businesses are familiar with your needs and are easily accessible. These brokers may possibly charge more, so do your research. You'll be paying a considerable amount of money for insurance and you will have a long-term relationship with your insurance broker. So shop around. I personally have three different companies that I use for my liability insurance, worker's comp, and disability. It was the most cost-effective solution for my business. When I need certificates of proof of insurance, I do have to contact three different parties, but my system works fine for me. You will find your own balance as well. More on billing later, but your time to gather required paperwork should be included in your quote. If it takes you an hour total to get your broker the information they need to get your insurance certificates and then pass those along to the requester, the quote will have an item for "insurance requirements" on it at your hourly rate.

Expect liability insurance to cost between $100 and $500 per month. This depends on where your business is located, the type and amount of work you do, what you want insured, and a slew of other factors that you'll go over with your broker.

Umbrella policy can cost as much as the liability. Let's say you have a $1 million limit on your general liability insurance policy. Some high-end condo buildings or commercial properties will require you to have an umbrella policy over your liability insurance, generally of $5 million.

The same company writing you the general liability policy should write the excess liability insurance. Insurance companies offer umbrella policies to cover your existing insurance from $1 million to $10 million. Cross this bridge if and when your situation asks for it, but it is a good idea to inquire with your insurance company if they offer such policies and at what cost.

Expect worker's comp to cost around 10 percent of payroll. Let's say, for example, that you are the only employee and you expect to make $60,000 for the year. You will pay an estimated rate until they audit you at the end of the year and your rate is adjusted based on the real figure. At your above expected salary, you will have to pay worker's comp for yourself at a rate of about $500 per month. NY State has a minimum payroll rate of $32,000. Even if you make less, you'll still be paying worker's comp at the minimum rate. If you use subcontractors or independent contractors declared on 1099s, you'll also be paying worker's comp for them if they do not have their own policies. So when you calculate your labor rate, remember to add about 10 percent to account for the real cost of labor after adding worker's comp.

Disability insurance should cost around $100 for the entire year. You got a break on this one, so let's not contemplate further.

These are ballpark numbers, so check with your local providers for more accurate rates to fit your specific business.

Bonds are another type of insurance that you'll come across. Generally, the agency issuing your contractor's license will require you to have bonding. These bonds ensure that contractors follow all applicable licensing laws and regulations, protect the consumers from damages done by the license holder or employees seeking unpaid wages. This bond is mandatory and usually around $20,000. Since they are so common, they are easy to obtain and cost only a few hundred dollars for the period your license is issued (two years in NY).

There are several other types of bonds that you may encounter. Usually all government projects will require bonding. Some commercial work and other new construction will need various bonds. These bonds could have face values in the millions of dollars and are difficult to obtain, especially for new businesses. You will need audited financial statements and will need to pass many other rigorous checks to qualify. These bonds include: bid, performance, payment, maintenance, and other bonds. These bonds are calculated as a percentage of the bid price, are obtained on a per-need basis, and will be calculated in the estimate. The small home improvement remodeler may never need these, so explaining these bonds is beyond the scope of this book. For further information, go to a surety bond broker.

Looking at the above figures and adding other fixed costs such as car insurance and loan payments, maintenance on equipment, phone and Internet bill, website hosting costs, contractor licensing fees, storage costs, and other costs that tend to add up quickly, the picture starts getting grim. You could be spending well over $1,000 a month just to stay in business, whether you're working or not. I will make this point repeatedly throughout the book: cost control is essential. You must think and act like a savvy CEO. Cut out all unnecessary expenses and pay the lowest rate possible for things you cannot live without.

BRANDING

I was told in business school that the best way to copyright or trademark a name is by using it. So go ahead and use your business name frequently and put it on everything pertaining to it. You gain rights and protection just by using it. Still, you want to make sure it is unique enough that you will not confuse your customers, or steal someone else's customers. Calling your home improvement services operation "Home Depot" might be pushing several envelopes. So do a quick search online for the name you want. A unique name will also get you the website domain you want and other advantages. I do know a contractor who named his construction company Dewalt and used a yellow color scheme. That is just cheesy and I cannot take his business seriously. If his creativity and foresight are that limited, how limited will his construction abilities be? Don't be that guy! If your last name is Hilti, consider using your first or middle name or full name for your business, and stay away from the color red.

Once you have figured out what you would like your business to be called, stick to it. Do not change the name or use different names on different advertising. Your business name, and tag line if you choose to have one, should always be the same, using the same font and color scheme.

Whether it is displayed on your website, your business cards, a banner, flyers, invoices and bid proposals, t-shirts, or a giant yard sign, your business identity should always be consistent and easily recognizable.

As you run your business and constantly seek improvement and growth, you should always ask yourself the following questions:

- Who am I, and what am I exceptionally good at?
- Who are my customers and what do they want?
- What do I want my reputation to be, and how do I work toward it?

These questions may not have answers in the beginning of your career, but should start forming as you get more jobs and interact with clients. You should always evaluate your jobs after completion and ask for feedback from both clients and your crew. You want to make sure you pay attention to the things your clients care most about. Not only will your reputation speak for itself, but also you will have selling points when trying to book the next job.

Market research is important as you find your business identity. Constantly seeking criticism will help you evolve. You want to provide exactly the service being requested. You want to be good at the things that matter. You first have to understand what matters to the person writing the check.

There are many stereotypes most homeowners associate with contractors. They are either messy, dishonest, never finish on time, sloppy, cost more than they promised, the quality of work is questionable, or they are a downright annoyance to work with. Make sure you understand the things that scare your customers and communicate why and how you will alleviate those concerns. You want to be different than all those other nightmare contractors. You want your clients to see that. You want to gain their trust. Most importantly, you must deliver on your promises. As Warren Buffett wisely put it, "It takes twenty years to build a reputation and five minutes to ruin it. If you think about that, you'll do things differently."

All of the above is very important because your brand is not just about your look on paper, but is really who you are. How you conduct business is more important than a fancy logo. At the end of the day, you live on with your character.

Branding highlights:

Name your business

Pick a name that is not too long or difficult to read and pronounce. Someone driving by a sign should easily pick up the name printed on it. Make the name relevant in your trade. Do not try to be grand and come up with cheesy gimmicks in your name. If your name is John and you are an electrician, "John's Electrical Corporation" or "John Electric" could both be solid names. A name like "John's Worldwide Electrical Enterprises" will have your customers asking a lot of questions.

Design your look

Choose fonts, colors, and sizing that are not too designed and eccentric. You are a professional. A bold and strong look commands respect. A tacky design looks cheap and shows immaturity. Be careful with tag lines and logos. Your tag line may inspire someone to call you or may just have them thinking you are not as professional as you think you are. Let's say you are a demolition contractor, and you own "Joe's Demolition Crew." You may come up with the tag line "Fast, Clean, and Under Budget." This straightforward line conveys that your pricing is reasonable while you work fast in a clean environment. You may think a tag like "We have bigger wrecking balls" might catch more eyes. Funny lines sometimes work but sometimes do more damage. You may certainly offend some folks. This particular line might just work for a demolition contractor, where precise work isn't key and shows a relaxed attitude and maybe someone easy to work with. However, using a funny dual-meaning line if you are an interior designer might backfire. Remember you want to be proud of your image and you want it to represent you and what your business stands for. The same goes for your logo. You are not opening the next .com startup or selling some new gadget that we all can't live without. You put nails in walls. Is a fancy digital image appropriate? You're the boss, so you make the call. Most often, less is more. A clean look with less clutter always wins. When in doubt, go back to market research and ask all your friends what they think. For most professionals, a solid name with no cheap tag line and no custom logo works wonderfully. Sometimes, small bullets under the name are appropriate:

LOMBARDI & SONS CARPENTRY
- FRAMING
- FINISH CARPENTRY
- CUSTOM WOODWORK

You have given your customers a classic way of seeing your name with a snapshot of what you do. Your brand is entirely yours to choose and live by.

Use a website

We live in the twenty-first century. You will not get too far without a website. I do not think I have to beat this point to death. Everyone will ask to see your work and they will ask if you have a website. This is the world we live in, welcome to it and start fitting in! More on website design in the next chapter.

Use your brand

As mentioned before, stick with what you decide and be consistent. Use your image everywhere. Every time you send a bid, or an invoice, or a proposal, or a letter, or any communication coming from your business, your brand needs to be prevalent. Your business name, tag line and logo if you have them, your phone number and address, your website, and your contractor's license should all be used. You can easily design them yourself and save them as templates in Microsoft Word or Excel.

A note about text editing software: After you write anything in your favorite software, do not save and email a client that document format. Always convert it to PDF.

Here is why: Your client may not have your software and may not be able to open it; your client's software might open it in a different format with your text aligned wrong, giving your file a messy look; you do not want people editing what you wrote, especially for contracts and other legally binding material. For a professional look that no one can mess with, convert and send everything in PDF.

Spread the word about your business. Tell everyone. Talk freely and excitingly about what you are doing. Give examples of your work. Incorporate current challenges and how you are solving them in your conversations. Be ready to sell yourself while waiting in line at the grocery store. Understand the key things that make your business better and how you can offer solutions. Deliver your message clearly with credibility to motivate your audience into becoming a client. Always carry some business cards and give them out if appropriate. You will evolve and so will your branding. It is a journey of self-discovery. Know and live your branding message. We should all constantly and tirelessly seek improvement.

Chapter 3
Where Do I Start?

Sell a man a fish, he eats for a day;
teach a man how to fish, you ruin a
wonderful business opportunity.
—Karl Marx

Most contractors are smart individuals. They are problem solvers. They multitask and think in parallel universes. They get paid to worry. The clients write the checks, so they don't want to worry. The workers make a daily check regardless of the job, so they don't worry much either. All the responsibility ultimately falls on the contractor. They deal with all the stress from the crew, the work itself, the business (or lack thereof), and the client. They fix what stumbles them, make it look good and move on. They don't really have any other choice. They have to. The ones who cannot handle the pressure, never make it in this industry.

If you think you're about to enter a life of leisure, dreaming of being cell-phoned while you're on a beach with news of how great your project is running, you should toss this book and go get a job. I do have the feeling you know better. For your sake, I hope you're one of us, mentally tough and ready for war. I remember the Soldier's Creed while in the US Army. Some of it said, "I am disciplined, physically and mentally tough, trained and proficient in my warrior tasks and drills . . . I am an expert and I am a professional." You certainly do not have to be a soldier to relate. If you replace the metaphor "warrior" with "contractor," you have the contractor's creed. I could give so many other relatable examples from the Army, but my point is that you're not just building a business, you're building Noah's Ark, as you're about to embark on quite the storm.

There is a really nice quote from Farrah Gray: "Build your own dreams, or someone else will hire you to build theirs." The reason you are reading this book and considering starting your own construction business is because you think you can do it better than the guy you'd be working for. Most contractors become their own boss after working in their trade for some time and realizing they can just work for themselves instead. The guys who keep their jobs and work for someone else have a more limited mental capacity. You'll soon realize that the problem solvers aren't really always the guys working for you. If they were you, they'd be working for themselves.

You decided you want to work for yourself as a contractor. Here are some options to get started in the business.

TRANSITION FROM WORKING FOR SOMEONE ELSE TO WORKING FOR YOURSELF

Most contractors go off on their own after they have gained enough experience in the construction field. This is a very good way to become knowledgeable and learn the ropes without risking your own skin. If you've been working in your trade for some time and feel you can manage jobs on your own, then maybe now is a good time to take the independent leap. If you are just now getting started in construction, then you have some work to do before you can be boss. You should have a few years' experience with general construction. Look for a job as a helper to get your foot in the door. Try to gain as much exposure to as many areas as possible. You must have a "can do" attitude. Get your hands in carpentry, plumbing, electrical, tiling, flooring, and finish work. If you work on houses, you'll also want to see how cement work is done, roofing, and how other structural and outdoor work is performed. The idea at this point is to get educated. You will make up your mind later on what direction your business should take. The more you see, the more you understand how to solve problems that inevitably come up. You may never need to be an expert in any area, but you should have a great understanding of what it takes to get the job done right. Understand the tools and techniques involved. Over time, try to gain more and more responsibility. If you are proficient in a certain task, ask to do it yourself without help or supervision. You need to prove yourself. Remember, you must perform. If you promised to do something, you must do it right. Do not cut corners or improvise things you do not fully understand. Never be afraid to ask questions or seek advice. It is very infuriating for your boss to come check your work only to see you messed it up or did something entirely wrong. You will lose credibility fast and your responsibilities will be taken away. You must stand on your own and be dependable. By taking these steps you will understand firsthand what the workers go through, how jobs get completed, and how to manage others. You will learn from the good and you will see things you want to avoid when you are boss. There is nothing more valuable to understanding how your business should run on a day-to-day basis than working in it yourself. You'll most likely start your business working side by side with your crew. Most of us did and many small contractors are hands-on their entire careers. Working with your guys will get you accustomed with how long tasks should take and will make you a better estimator and manager. You must understand your business before you can run your business. At the core of it all, before you can run an efficient back office, the jobs must be completed timely and precisely. Do not ever forget the role of your business.

Get educated

You could seek formal education. For most small contractors, all the education they ever need is provided on the job. Running a solid multidimensional business, however, requires a vast understanding of management, marketing strategies, finance and accounting, business administration, and an entrepreneurial spirit. If you intend on working for a bigger construction corporation and eventually becoming a general contractor, you should pursue a degree in construction management. Certificates and technical diplomas are also widely available and worth looking into. As mentioned in earlier chapters, I am a firm believer in constantly seeking improvement and taking advantage of all opportunities that could teach you something. Learning should be a lifelong goal. On top of your hands-on education, you should consider formal education that makes you more knowledgeable and credible.

If you are in a specific trade like an electrician, and you want an electrical contractor business, then you know exactly what business you'll be running. If you feel that painting is suitable for you and you know you can manage that business and excel at it, then follow that path. If you do not quite know which way you want your business to go, maybe you should let your customers dictate. You'll find your footing once you start booking jobs. Or maybe you do not want to specialize and you are a jack-of-all-trades. For more of a general contractor type, you should plan on making strong connections with subcontractors. You will be managing the jobs but the work is being delegated and performed by subcontractors. These subcontractors are efficient at their tasks since they do them day-in and day-out. They may not need supervision, but will certainly need to be kept accountable and on schedule. Since you are the manager that puts all the subcontractors together to finish the puzzle,

49

you need to understand how to put the puzzle together and where to get the pieces from. You will be doing a lot of research on subcontractors. You need to start putting a list together of all different businesses in different fields. You will need to read reviews online if you haven't used them yet or get references. Recommendations are very helpful from your circle of contacts. Always reach out to those you know. You will need to find competent subs who will work with you on pricing and scheduling. The task can be daunting, and your first few projects will really test you. Things become more routine as you grow your business, but you'll still need to replace or update certain subs.

Regardless of all the knowledge you've acquired and experience you have doing certain tasks, you will most likely run across work that you have not yet done. You may have an idea of the way to go about it, but you are uncertain of the proper technique and materials used. The Internet is a wonderful place to start. In today's digital age, anything you can think of has been documented and shared online. You are not sure how to correctly build a shower pan? No problem. You can spend hours watching videos on YouTube. You can see good technique but also you'll find "experts" who are sometimes questionable. The good part is that you can see so many videos with explanations that you'll be able to see exactly what the excellent tricks have in common, and where you should do some more research. Watching a video should be a first step. Reading proper techniques on websites as well as shifting through blogs that bring up potential pitfalls should be next.

People post issues they have had with certain work and you may find answers from everyone pitching in with solutions. Remember to always consider the sources. In general, the more common ground and similar solutions you find, the safer it is that you have the right answer. There are also many well-written and illustrated books on just about anything pertaining to construction. They can be found in Lowes and Home Depot as well as on Amazon. They are a great reference to have for go-to last-minute practical solutions.

As with most life challenges, you generally should not go at it alone. There is no need to reinvent the wheel. There is great value in having a mentor. You should try to befriend someone in your industry with whom you can share wisdom. Mentorship is a mutually beneficial relationship and when you are in a position to be helpful, you should be. Having someone to share your struggles with is invaluable. Sometimes, just speaking your issue and having them repeat it in a different "language" makes you think of solutions. Having another contractor that you can bounce ideas off is very important. I believe most people are good-hearted and well intentioned. Given the opportunity to do some good in our community, we all lend a helping hand. Consider carefully who your mentors are, and regardless of your status or achievements, you should have an adviser and someone who can advise in time of need.

Subcontractors are also a great learning source. There are times when you do not have the knowledge, resources, or time to do a certain part of the project. So you call in another pro who specializes in that work. It may cost you more money than if you did it yourself, but you are essentially paying for your education. You will be getting a first-hand demonstration of how to properly do the job. You'll be watching their technique, the tools and materials they use, and other tricks. They'll ask you questions, so you'll ask them questions as well. They made some money and you got a lesson. This relationship may continue as you sometimes realize they are more efficient and you'd rather avoid the headaches of doing it yourself. There are certainly things I do not want to be experimenting with even after I watch a pro do it. I would certainly not trust my crew with moving a circuit panel in a home. Some things are better left to the guys who have the proper schooling. In general, you'll learn a lot from other contractors you hire or those who hire you for a job. If you're on a larger job and you're only doing the sheetrock, you'll of course naturally look around to see how other things are done by other pros.

Your crew is another source of expertise. Some of the guys you'll be hiring have a ton of experience. Most guys probably won't. But even the guys who aren't quite so proficient have probably worked for other contractors and have seen how problems get solved elsewhere. I always learn something from my guys, and I try to teach them what I know. It only makes your business more efficient. You have certain ways you do things, but they may have other ideas or tricks they have learned from past work.

Always be open to seeing other points of view. When giving a certain task to a worker, I often ask, "How would you go about building this?" I listen to his answer, then make adjustments and corrections if necessary. Sometimes I hear solutions I haven't even thought of. Communication and cohesion strengthen your crew. Sometimes you'll just hire a guy because he was recommended as an excellent tile guy. While you watch him work and help him, you'll learn the proper way to tile. Again, you have not only paid for his work, but you bought yourself education.

I will make one last comment about learning. Read the owner's manual of your tools. I know, you are handy and can certainly figure out any tool on your own; however, this isn't just about how to recharge the battery on your drill. Most tools will give you insider tips on specific applications. You will be pleasantly surprised to learn tricks about wood grain from your table saw, cement applications from your rotary hammer, and many nuggets of wisdom you never knew belong in an owner's manual. Plus, you will discover cooler ways to use your tools and you will stay fresh on the entire tool lingo.

MARKETING

Marketing is often wrongly identified with only selling. While selling and promoting your business are major parts of marketing, more important is meeting customer needs and wants. If you sell rain jackets and direct all your marketing efforts to communities in Libya, you won't last long in business. Last I checked, Libya is on the Sahara desert; you would probably do better if you were selling water.

In your marketing efforts you must:

- Define your business
- Identify who your customers are
- Understand what your customers need
- Come up with a plan to give your customers what they want

Remember you are selling yourself all the time. It is about projecting a trustworthy image of your business. Your marketing plan should meet your customer's needs to bring you profitability. Now that you are a business owner, you have to promote your business not just through well-defined advertising campaigns, but also instinctively on a daily basis. You are the face of your business, so you must be ready to speak excitingly about your work at any time. You never know who may have a friend in need of a professional like you.

Your number one marketing tool is your work. Think back to the Warren Buffet quote about your reputation. It takes a long time to build it and it speaks for itself. Always think about your reputation and work accordingly. Work in someone else's home as if it was your own. Demand the highest quality standards from your crew. Always instill a high-end mentality in your workforce and never rush or cut corners. Do quality work, and you will keep loyal repeat customers.

Your current customers have friends and family who will see your work when they visit the home. What do you want them saying about the work? What do you want those friends telling their extended network of acquaintances? Your work will market itself if you pay attention. Besides the finished product, their experience dealing with you will also play a major role. Were you honest, timely, clean, responsive, caring, fair, professional, and easy to work with? If you were, you might not only get their future projects, but their recommendation. Work on being a perfectionist when it comes to the job and work on being respectful when it comes to the client. You will spend a lot less time looking for work when the work comes to you.

Your first go-to people in your network are brokers and designers. They interact with your prospect clients on a daily basis and have established some trust already. Interior designers need contractors to make their design ideas come to life. They are always looking for dependable and responsive contractors. Brokers help homeowners get into a new place and can recommend contractors before the clients even know they need your business. Make sure you become friendly with brokers and designers. They must trust you first before they can recommend you. When you meet them, offer to do small jobs to get your foot in the door. Since you won't have much work to show you can start by taking pictures of small jobs you have done in the past. Even if you work for another contractor, you should photograph the work you do. Keep all pictures of any construction work you do. They will be your first proof of your workmanship. Establishing relationships with other professionals will get you work without having to spend any advertising dollars. Also, when selling yourself to another professional who understands your business is easier than to a homeowner who is suspiciously afraid of contractors to begin with. Work on making lasting connections with brokers, architects, decorators, property managers, engineers, designers, and other contractors who may use you as a sub.

Paper-related advertising vehicles

You should have a stack of business cards on your person at all times. When you give them out, you should talk about your business and give interesting examples of projects. You need to give them out to everyone you encounter, without being pushy and obnoxious. Think about trying to pick someone up at a bar. What do you think happens when you just walk up to someone and hand him or her your card? Your business card ends up in the garbage. Your goal is to get *their* number. If you can score their phone number, you have already established some sort of rapport and built just enough trust to get the number. This goes for any first-time interaction. You want to get the other person interested in you and your business. If you can get their contact information, then you have an open door to walk in and secure the job. Try to get their info first, and then give them your business card as a reference. Afterwards, when you send them an email, call them, text them, or send them something in the mail, they'll remember you.

Every job you are working on creates an opportunity to meet the neighbors and set yourself up for future work. You should ask the homeowner if you could put out a lawn sign. Explain how this will warn the neighbors of any noise and disruptive activity while you're working. Make a large sign that can be read while driving by. Remember to include your designed branding material. In addition, you can write what is being done at that location. For example, you could say RESTORATION WORK IN PROGRESS, BY LENNY'S CONSTRUCTION CORPORATION. A nice tag line on your lawn sign to inspire the neighbors could be BEAUTIFYING YOUR NEIGHBORHOOD. Everyone is curious of what the neighbors are doing. Grab their attention with a lawn sign. You need to go even further in making yourself known. You should have a flyer made or a short letter to pass out to all the neighbors. Put them in their mailboxes or slide them under the door. If you have the opportunity to meet the neighbors, introduce yourself in person. You need to apologize for any noise or inconvenience caused during the renovation next door. This will show them how professional and caring you are, and will put your business on their map. Also ask that if any issues arise they contact you directly. If there are problems, you want to be notified first so you can handle them before they get out of hand or passed around to others. Of course, your branding material will be highlighted on the flyer. A word of caution on such marketing strategy: you and your crew must be extra careful while working in the neighborhood or condo building. You must live up to the expectation you just set for yourself. Be sloppy, and everyone who now has your marketing material will remember you by that.

Other advertising freebees may or may not work. Items that you give out such as pens, beach balls, squeeze balls, and mouse pads are expensive and tend to not have much effect on the end user. They always seem to be given to the kids or the pets to play with, and end up in the garbage at some point down the road. I would stay away from spending your money on memorabilia that no one pays attention to. The one item that I think could make sense is a magnet. Most families have things to pin on their refrigerators and can always use an extra magnet. The magnet will have your information readily available when the time comes to fix something in the house. Another free item that any homeowner would hold onto would be a multi-bit screwdriver. Everyone needs a screwdriver at one point or another. One of those convenient interchangeable-bit screwdrivers would come in handy and would have your name all over it. The homeowner will be reminded of how accessible you are every time that screwdriver comes to hand. Again, these giveaway items can be expensive and may not say much about you as a contractor, so use your judgment and always err on the side of cost control. T-shirts for you and your crew are definitely well received. They show your clients that you run a serious business. Oftentimes, your guys will even wear them on other jobs, giving your name more visibility and free publicity.

Advertising online

We live in a fast-paced, digital age. People Google each other before going on a date. Employers will check your Facebook page before the interview. Folks will read reviews on Yelp before they eat at a restaurant. We all read reviews on Amazon about products we consider purchasing. Everything we want to know about something public is easily found online. You can bet your clients will expect to put your name in their smart phones and pull up your work. If no one can ever find any public information about you, they might get worried. If everyone can find drunken pictures of you at frat parties, you should be worried. Share what you must on the Internet, but be mindful of things you do not want representing you professionally. Having an online presence is essential for your customers to find and familiarize themselves with you and, if done right, also draws in new business.

Use social media to your advantage. My mother got her first email account in her late 50s. She is now on Facebook. If even her generation is getting plugged into social media, you should have no problem mastering it. Your friends, their friends and your potential clients, are using apps like Facebook, Instagram, Twitter, Google+, Pinterest, YouTube, and LinkedIn. You should have a presence on all of them and be active on most. Having a fan page for your business that can be shared with friends makes your network of acquaintances, and potential clients, larger. Posting pictures and videos of your work is essential to people seeing your craftsmanship quality and to show that you stay busy and current. Doing tool reviews or "how to" videos on your YouTube channel is a great way to gain credibility. You should give people a reason to follow you. Even in the beginning, do not add friends when you have an empty page. Provide interesting information about your industry, post pictures of your work and your crew, and videos of happy customers. Pay attention to your followers and online friends. Comment on their posts. Pay attention to their birthdays. Build a relationship and create a circle of trust. Do not just pitch your services. Networking is not a sprint but a marathon. It is about building relationships that are mutual and pay off over time.

Become a trusted expert source by reviewing professional tools or answering construction questions. You can start with Home Depot. Review tools you own and always use your business name. They also have a section where DYIers ask questions on home renovation projects. You should answer them from your professional point of view. Please be very careful not to give misleading information. Remember your reputation and do not answer things you are not an expert in. Always save the links to any online activity and post them on your social media. Staying active online requires commitment on your part, as you must make time for all aspects of your business. Especially in the beginning of your career, you will have to work overtime to get your business recognized.

Having a website is a must. It is your portfolio. Your customers will want to review your work on their own time and feel comfortable with what you have to offer. It is important that you have a well-designed professional-looking website that is easy to navigate while offering enough content without feeling busy. You need to make a good impression in all aspects of your business. You are an overall professional, not a sloppy handyman. Project that image in everything pertaining to your company.

Your first step is to secure a website name. Domain name registration costs about $15 per year. You may not find the exact name you want but you should try searching variations until something that makes sense for your business comes up. Getting the name you want on a ".com" extension is ideal. I would stay away from other extensions as people are not familiar with them and tend to just type in ".com" as a default. There are new extensions that you may buy ending in: ".builders," ".construction," ".contractors," ".repair," ".house," ".company," ".services," ".trade." One of these can fit nicely at the end of your business name, and it could give you the perfect website name. For example, if you start a business with your partner and decide to use your names, Joe and Jane Construction, a perfect website could be "Joe&Jane.partners." Again, be mindful that most folks out there have no clue these extensions exist and may be puzzled on how to enter your website name in their computer. Also, your email will now have your website attached to it, so make sure the email is easy to read. If you decide to use one of the above, make it obvious in your branding design that this is the website with a clear "dot" separating the extension, and possibly the use of "www" in the beginning. I am presuming that in the future we will all be able to use any extension we want. Until then, odd extensions are not common.

The next step is to get online hosting for your website. Your pages have to sit on a server for everyone to access. You do not want the hassle of having a server installed in your home or office, so you need to rent space on a public one. Your emails also need to be stored somewhere. You will no longer be using email services like Hotmail or Yahoo; you are not in high school anymore. You own a business and your email needs to reflect your own website. Basic website hosting costs about $50 for the year. Many providers offer website design services as part of a hosting package. This simple solution may work for you, but analyze the long-term cost of paying a monthly fee for your website design and look at how professional it really is. Buying a website with several pages already designed costs about $50. This is a one-time fee and you can choose from thousands of website designs you may like. Afterward, you can tweak the look and add your own content if you are computer savvy.

After you have taken care of the backbone of your online existence, you need to start designing your website. Look online for websites you admire and try getting inspired from their look. You should have pictures showing your best work right on the main page. There should be a link from those pictures to a separate page that showcases more details of your work. Include a contact page with all your branding material and a simple contact form, and have a page that talks about your company and the services you provide. This is all that your professional website needs. A simple and clean, 4 main page site to include: a home page, a projects page, an about page, and a contact page. Include a link on every page that says "free estimates," taking the user to the contact form. Have links to all your social media clearly visible. It is very important to include the fact that you are licensed, insured, and bonded, and write in your license number. This reminds clients they are dealing with a professional. Have your license be stationary, as part of your header or footer, so it is clearly visible anywhere on the site. Include all trade organizations and certificates you have. Think about your resume and neatly incorporate all that information into your website. Do not use too many colors or too many font types. You want a clean and neat look, just the way you would look on a job in their home. Avoid using too many words and long paragraphs; instead, try using bullet points for quick references. On the about page, write in your personal story. People want to make a personal connection and feel they can put their trust in one individual. Companies that talk only about "we" seem too large to pay attention to an individual homeowner's personalized needs. Make it personal without giving too much detail into your private life. Post pictures of you and your crew smiling and happily working. This will show the prospective client that you guys love what you do, and it will offer a glimpse into what is to come for them. Make your website viewer comfortable with your business and get them to like and trust you. Make them feel secure having you in their home. Your website needs to be warm and welcoming, not cold and rigid. If you love your children or your pets, you may just have something in common with most of your clients. Show them how loving and caring you are and they will relate.

There are many companies out there that will try to sell you their optimization services with promises of getting you top ranks on Google searches. Save your money as these services do nothing but get you frustrated and waste your time. Google has a proprietary way of ranking websites that constantly changes. You have a close to zero chance of getting to the top when someone searches for a contractor. If you live and work in a small town or region, then chances are you are one of a few contractors and you most likely will get noticed anyway on the local search. Pay-per-click advertising is available from Google AdWords. These ads can be featured at the top of any searches, but will cost you quite a pretty penny. You can spend thousands on advertising on AdWords and get no business coming from it. You must understand the right key words your customers are searching for and what words make them determined to buy right away rather than just shop around. This is quite an expensive advertising campaign. If you do decide to try it, set small budgets and monitor the clicks closely.

Writing a blog or article online about your industry is beneficial but takes a lot of precious time. If you constantly update your site with articles about "how to" or "mistakes to avoid" or other construction-related ideas for consumers, you may get top rankings when people search for construction companies in your area. This is time consuming and since you're at the beginning of the road, you need to focus on many other things.

Other ways to consider drumming up some business online are through classified ads and lead providers. A very popular website for handyman type of work is Craigslist. Many small jobs get posted on there by homeowners or tenants. I would try it first but be warned that these are small jobs coming from people who will nickel and dime you. They are also public and free, so many small contractors will bid for the same ad you are reading. It is worth a try, especially if you can't find work elsewhere. I have had an electrician tell me a horror story about how he finished a two-day electrical job for someone who gave him a check that bounced. Demand cash and some sort of deposit to cover you. There are all sorts of people out there and the last place you want to find yourself is in the middle of a legal battle. Most of the time it is not worth your time and effort. Make your customer sign a short paper agreement before you start that explains the work done and the charges. This will at least instill in their head that you are covering yourself legally in case they think about pulling a fast one. Hopefully this will deter them from being shifty. Angie's List is a system that requires the user to subscribe in order to avoid less serious clients. Advertising there gets good results for some contractors.

A wonderful way to get credible qualified leads is from websites that do exactly that. There are many sites out there that will get customers to input their project needs and send them a few contractors to bid on that job. It is free for the customer, but you, the contractor, have to pay a fee for the lead. Pretend you are a homeowner who is looking to get something done around the house. Do a quick search for a service you provide in your area. You will be led to websites that promise to send you several contractors to bid on your project. Those are the sites you want to target, since they will give you those local customers looking for your type of business. BuildZoom and HomeAdviser are reputable companies that specialize in matching projects with contractors. As with any advertising budget, keep track of your cost per job. For example, if you pay $50 per lead, and you book one out of four jobs you bid for, your cost to acquire that job is $200. Always keep track of your expenses, and in this case the cost per job will tell you if the advertising strategy makes sense for your business.

Very good promotion for your business is an article written about you. Try to do charity work once in awhile. Especially if you have the time, contribute to the local community. Help out those in need or help the local government. Be sure to get the press involved. A small article about how your caring construction company is helping the community carries a ton of weight. Remember to post those links on your web page and social media pages. You will feel good about yourself and you will create an immaculate image for your business.

In today's fast-paced and technologically advanced world, we forget about the human touch. How many Christmas cards did you get last year and how many did you send out? When was the last time you wrote a letter and put it in the mail? Do you remember the last time you got a note in the mail that wasn't junk? Maybe you got an invite to a wedding. You saved that card, at least for some time, didn't you? Did you save the emailed invite? Most likely not. You save the card because it is personal. You treasure it for being rare and having some value to it. You should create that feeling in your business dealings. A simple handwritten note for a meeting, or for an apology, or after finishing a job goes a long way. You will most certainly stand out. You could buy your client a small gift at the end of the job, like a picture frame, or mail them a gift card from Dunkin' Donuts. Pay attention to what they like and the places they go to daily. Send a small value of $20 or so with a thank you note and a reminder for a referral. They will feel obligated to get you a referral because now they owe you. When they use that gift card to get coffee in the morning, they might just tell someone in the store about you and what a wonderful contractor you are.

You may consider becoming a service provider for Lowe's or Home Depot. While this may be difficult in the early stages of your business, becoming an installer for one of the big box stores can increase your workflow and expand your horizons to new customers. They have a thorough application process that can be started right on their websites.

Reviews are very important. Always ask for reviews from your happy customers. Even where the job ran into difficulties, after you solved the issues those customers should be pleased with the way you handled yourself. A review that shows how you mastered a tough situation is constructive. Customers considering hiring you will feel they know you better from reading reviews. Ask your customers to post them on your profile on sites like BuildZoom, or copy and paste them on your website. Keep them updated so your work looks current.

One thing to really be conscious of: Before you grow your business, make sure you can handle it. Once the leads start pouring in and the phone keeps ringing for job estimates, make sure you have a system in place to handle all the work coming your way. If you work with only one other guy, one project at a time, you are not ready for growth.

You need to set yourself up for what's coming. You personally will need the time off the job site to go see prospect clients, draft up bid proposals, and organize jobs to come. Then, you will need a crew of guys to do the actual work, and possibly several crews and several sets of tools if you take on more work. Be very careful to not bite off more than you can chew.

Chapter 4
Getting the Job Done

*Whenever you are asked if you can
do a job, tell 'em, "Certainly I can!"
Then get busy and find out how to do it.*
—Theodore Roosevelt

Entrepreneurs have a tough time saying no. They are hungry to grow and see their ventures thrive. They love challenges and enjoy solving problems. This attitude is partly the reason for their ultimate success. Most contractors I know, myself included, find it very difficult to turn down business. Even when we have our hands full, are working overtime with a crew stretched thin, and possibly have never done what the client is asking for, we still take that job. We do it for our egos, or to gain experience, or to prove something to ourselves or others, or maybe for bragging rights, or the opportunity cost, or possibly to challenge ourselves, or for the fear of losing a repeat customer, and certainly we do it in hopes of a bigger payday. Small business owners know how to swiftly hustle. You would have to, in order to successfully stay in business. You have to be willing to constantly learn and make progress. You will find yourself in tight corners at times. Getting through those testing moments successfully will only make you a stronger business owner. When the time comes when you are asked if you can do a certain job, before you can even consider what resources to use, you may find yourself answering yes. You nod your head confidently while saying "absolutely," then you turn around and walk away puzzled on how you are going to get this done. You reassuringly congratulate yourself; you at least got the job and not your competitor. Now go get it done!

I started my construction business in Manhattan while living in a studio apartment. I have certainly come a long way, but I am still amazed at how efficient I was back then. While I have no interest in reliving those times, I am certain that anyone can start a new contracting business on a tight budget with limited space. Humble beginnings make you appreciate later triumphs that much more. I was able to walk into a multi-million-dollar apartment and run confidently over construction steps with high-profile clients. Little did they know that at home, my computer desk was right at the foot of my bed, that bed had all my office paperwork tucked under it, my closets hid tools under stacks of clothes, and pots and pans shared space with circular saws and routers under the kitchen sink. My coffeemaker was on top of the wet saw and I had buckets of paint and bags of cement under the fish tank. My dad would mumble "there is wood all over this kid's apartment ..." as he tripped over 2 x 4s. "You live with wood inside your house!" Mom would have to duck her head in the kitchen to dodge the bike hanging from the ceiling. I loved when my parents visited and, as rough as it seemed, I was actually proud of what I was doing. I lived alone but found great company in my tools. I was starting out and was excited to build a future. If running a construction business in the heart of New York City from a studio apartment turned successful, you should be smiling at your prospects. Try bringing a date home and explaining why they have to climb over a shop-vac, extension poles, and caulking guns to get around the apartment. Hopefully they have a sense of humor.

THE TOOLS FOR THE JOB

As a contractor, you will need quite an arsenal of tools to aid in your daily work. Things like your vehicle, a computer, and your phone are indispensable, while the massive inventory of power and hand tools for completing jobs is inevitable. As time goes on, you will find the need for tools you never even thought about. You will need the capital to acquire them and the storage space to hoard them. If you are a specialized contractor like an electrician, plumber, roofer, finish carpenter, or demolition contractor, you may only have the need for a handful of tools you use everyday. You keep them all in your van, and they are easy to organize. On the other hand, if you are like me and want to get your nose in every aspect of construction, you will end up with assets that rival the equity in your home.

Your vehicle is your primary resource for getting to work and bringing along your tools and supplies. I find it indispensable and thank heavens for it almost every day. It has brought me more headaches than I can count, but in the end, its benefits far outweigh its small annoyances. If you live and work in most of the United States, you may actually not have a choice but to own a vehicle. Only in major urban areas like Boston and New York City do people get by without a car. One of my very close friends, a successful contractor for the past two decades, has never owned a car. Living and running his business in Manhattan, he actually has no intention to ever complicate his life by having wheels. It is a different business model, but it works extremely well for him and other small contractors in big cities. He has streamlined his resources for pickups and deliveries in a way that enables him to delegate all transportation needed to the appropriate parties. It costs him more money to pay someone to do all the things he could do with a truck, but he saves considerably on gas, tolls and parking, car payments and maintenance, and tickets and insurance. In the end, depending on the volume of work, it may just end up being an equal cost structure but may save him some headaches that I'd endure for having a truck in the city.

If you live in a major city, you may choose to try this approach. You can have your supplies delivered by your local hardware store. You will find that many suppliers offer free delivery if you bring them enough volume. You can easily develop a relationship with a carting company to take all your construction debris. There are many guys with a van who will take all your trash and bring it to the dumpsite. You can find them on Craigslist, local ads, Yellow Pages, simple Google searches, and many times sitting in the parking lot at Home Depot waiting for a delivery job. Make several calls and shop around for the best price. Once you establish a relationship with a local delivery guy, you will get faster service and better pricing. Your network is very important in running a construction business efficiently over time, and it will reward you in times of crises. The first calls you should make whenever looking to find someone are to the ones in your network. Your carpenter may have a friend who owns a van and does deliveries, your plumber may know a guy who is looking for work and would be happy to do some general labor, and so on. I find that I get most of my answers and needs filled from inside my network. For many reasons, referrals are much safer than calling strangers. This is also why it is important to be friendly with other contractors. Not only will you be able to bounce ideas off each other in times of uncertainty but they may have an extended network of contacts you may be able to tap into. Your contractor friend may need a demolition sub and you have just the right contact to give him. When you need a cement contractor, he may just have the number for the guys who paved a driveway for him. Everyone wins! I found that even my competitors sometimes ask for my help. I gladly offer it, knowing they are now indebted to me, and I may need a favor soon.

If you end up being a more traditional contractor, you will need a vehicle. Someday, when your business is large enough, you will probably need more than one, but to start with, you should have a van or a truck. They each have advantages and drawbacks over the other.

A truck is a masculine and powerful look synonymous with an endearing American way of life. This can be your only vehicle, as you may use it for work and to go to the beach without looking like you are hiding something inside a dark and shady van. Trucks are easy to load and unload from all sides and offer enormous space for large and odd-sized materials. Unfortunately, they are a poor choice for rainy days as all your tools and materials will inevitably take a shower on the open bed. You may choose to put a cap on your truck, but you would lose the flexibility of hauling large items and ease of loading, plus your full sheets of sheetrock or plywood will still be exposed to the elements due to their size. Trucks also offer very poor safety options for your belongings. You cannot leave your tools in your truck overnight, or even materials on the back of the bed while you are on the job. In some small and honest communities, you may be trusting enough to feel that no one would help themselves to expensive equipments sitting on the back of an unsupervised truck. For the rest of us working in most of the real world, the truth is slightly jaded. I have had scrap metal stolen from the back of my truck. Yes, it was garbage, but someone could make a buck turning it in to the scrap yard. In the big cities, you may come back to your vehicle and find a rock had gone through your window if you left a tool inside.

A van is a very practical vehicle that will offer the safety of locking belongings inside and driving to jobs through bad weather. A van, however, is harder to load and unload as you only have the back doors to work through. Really large and odd-shaped items would also never fit inside a van. Try moving even a small tree in a van, while keeping it alive. Vans are also impractical for personal use. I doubt you would want to take your date to dinner in your work van. Driving a van to your high school reunion may give a new meaning to the class clown.

Whether you get a van or a truck is the choice you should make based on your personal situation for intended use. Advertising your company on the side of your vehicle is common practice for contractors. Keep in mind there are some key considerations about tagging your vehicle. Most states require you to have a commercial license plate if the vehicle is used for commercial purposes. A work truck branded with your contractor info is considered a commercial vehicle. In major cities, having commercial plates can often get you parking outside of buildings you are working in, but you may have a hard time parking in lots intended for passenger class vehicles. Commercial vehicles can also only travel on certain roads. If you intend on having only one vehicle for both work and personal use, I would stay away from commercial plates. You will avoid major head-aches and paying for expensive tickets.

Your office is a tool you cannot forget, nor would you want to, when running a business. Your office is your lifeline. You will spend considerable amounts of time in it running your company. At the core of your business, completing construction work is what your customers see, but you need a strong office backbone to support your setup. You will be writing proposals, sending out bids and invoices, run-ning payroll, doing accounting and finance, doing research on products and techniques, ordering materials, commu-nicating with clients in writing, and other design work that supports your construction work. You need a reliable computer or laptop. For it, you will need word process-ing software to design your letterhead, and write propos-als and invoices. You will also need spreadsheet software to run your books or set up bid calculations. Microsoft Word and Excel are the most commonly used programs, but they are expensive. Google offers the equivalent versions called Docs and Sheets. They are free and include free storage online and the ability to access your files from anywhere. You can save your file as a PDF to be Microsoft compatible. The only drawback is that you must be connected to the Internet to use the software since it is web based. Speaking of the Internet, do not forget to factor in your Internet and phone bills. They are a part of your office. Your phone is also a crucial component. I find myself doing work on it constantly. When on job sites, I will be communicating with clients through email or text, researching products or watching how-to videos, finding info on manufacturers' websites, finding the nearest hardware store, and talking on the phone relentlessly. My smart phone is my office on the go. The printer and some filing cabinets round off your basic office. Remember to account for any expenses that are related to your business. You will get all the stationery products such as paper, envelopes, or postage when you find the need for them. My philosophy has always been to err on the side of spending very little money up front on anything until you need to. Why stock up on all sorts of goodies only to sit there and wait for work while depleting your bank account? Buy what you need after you get the work and you have a check coming your way.

You most likely already own basic construction tools like a drill, some sort of a power saw, and other hand tools. I would start working with what you already own before upgrading. As mentioned before, only buy the tools you need when the job asks for them. If you are booking painting work, your plaster trowels, rollers, and brushes are all you need. Why buy sets of wrenches and a welding torch before you have any plumbing work? Only get the things you need for that job. Many specialty tools are available for rent. Things like a wet saw, jackhammer, belt sander, power drain snake, pressure washer, and other specialized tools you may not have a need for on a regular basis. These you can rent when the job calls for them. Before you do that, make a quick financial decision based on a simple calculation.

Let me give you an example. We work in many condo buildings that have concrete floors and ceilings. A rotary hammer drill costs about $60 a day to rent. The first time we needed it, we spent the money on the rental and expensed it to the job. Since I realized that we would be using this tool quite often, due to the concrete environment we would be working in, it made more sense to buy it. The $600 price tag of the tool is amortized in its first ten days of use. Since I now own it, I can get years of use out of it and never worry about returning it on time. If you mainly work in single-family wood-frame houses, this tool should just be rented for the occasional use. Just think about the type of work you would encounter and decide if it is beneficial to own or rent.

A word of caution on cheap tools: It is always easy to justify purchasing the cheap brands that seem to have identical features to their high price tag equivalents, especially in the beginning of your career. Maybe you have used them at home with great success for many years. First off, using them around your house on the weekends for the occasional DIY projects does not put enough stress on these tools. When they get used daily for hours on end, the differences start to be apparent. Mixing cement with your dad's old Black & Decker drill will burn out the motor in a heartbeat. The professional drills have higher quality brushes and bearings and will not burn when overheating. I remember when we were doing a flooring job and my cheap portable table saw burned out halfway through the job. My guys had to rip flooring with a circular saw and a jigsaw. It was embarrassing and frustrating, as well as costing me time and money. In my experience, these tools are just not meant for the constant abuse of a job site. Second, when you show up at a job with the same cheap drill that the homeowner also has in his garage, he will be wondering why he is paying you to do it. You run the risk of looking like an amateur by working with tools intended for consumers. If you are trying to step up to challenge the pros, you should have all the ingredients they use and then some. Tools like Milwaukee, Makita, Bosh, and DeWalt should be at the forefront of your armory. If you price your jobs correctly and run the work efficiently while making a handsome margin, maybe someday you can also own some Hilti tools. I am hopeful you will after reading this book.

Use the right tools for the right job. When I first started doing tile work, I used to remove existing tile with a pry bar and a hammer. Piece by piece, I would chisel by hand all the wall tile. The fun part came when removing the floor tile that was cemented to concrete subfloor. I am starting to sweat just thinking about it. A small demolition hammer costs about $500. The pry bar and hammer I already owned. So it was a no-brainer. I would spend nothing on a new tool and still get the job done. The job got done all right, but not without frustration, small injuries, and time better spent elsewhere. Instead of spending four hours on removing bathroom walls, then using a sledgehammer and grinder to take out high marks in the concrete subfloor, we now spend an hour doing the same job. Adding all those saved hours over time more than pays for the tool, not to mention that your company looks more professional and you will have a less frustrated crew. Sometimes it pays to work smarter not harder. Some of you may still try old techniques instead of embracing modern technology, and you may someday come to your own conflicting conclusions. Regardless of your task or situation, always consider the time and effort involved and make the right choice on what tools and techniques are appropriate.

Just as I emphasized above on proper professional tools, I will make the same case for materials. Let me speak from personal experience: when you use cheap, thin-set mortar to set marble or glass tile, the tile *will* fall off the wall. Just take my word for it! Let me make a side note while I am on this subject: Always supervise your crew or have a system of checks in place. Your heroic workers may unknowingly be using the wrong materials or techniques. Later, when it becomes obvious something was done wrong, not only will you lose credibility with your customers but you will be responsible for tearing it down and rebuilding it at *your* own cost. I keep mentioning cost cutting and that materials are big expenses for your business. Using higher quality materials, however, can render better results and save you time. You can sell your business on that premise as well. When marketing your services, make a point to let the customer know why you are using higher quality materials and why the other cheap contractor may not be so cheap in the long run. I will give you an example on grout. Almost always, contractors use cement grouts to finish tile installation.

As time goes by, the bathroom becomes revolting from the cracked grout, the grout color fading or yellowing, and mildew growth in the grout. We have all seen it and customers are horrified at the thought of their beautiful brand new bathroom turning that way. Most people have no clue there is any other choice. Well, there is. Epoxy grout. I explain the differences to my clients and almost always they go for the up sell. It is considerably more expensive and difficult to apply. That is where you would make a premium. This grout never changes color, never cracks, and never grows bacteria since it is virtually waterproof. It bonds harder than the tile itself and can be customized with cool features such as sparkles that reflect various light colors or glow-in-the dark for a kid's bathroom. All you have to do is research new products in your field and always try to differentiate yourself from competitors by having more options to deliver higher quality to your clients. Charles Darwin said, "It is not the strongest of the species that survives, nor the most intelligent, but the one most responsive to change" when he spoke about evolution. You should continuously evolve in your business. We recently made the switch from the old, all-purpose joint compound to a new gypsum product. Instead of having to skim coat an area twice and fill in air bubbles while mixing in plaster, we now do one coat with easier workability. The product is slightly more expensive but saves us time on repeated applications. This is money well spent that will come back to you in form of time spent and errors made, and it is a point of differentiation. Using black pipe instead of galvanized steel for plumbing may sound like a quick money-making shortcut to your plumber, but shouldn't be overlooked by you when inspecting the work. Always use the proper materials for that specific job, and try using better materials when possible. Your finished job is better and you can charge more for it.

Between all the tools and material leftovers, it is starting to look like you need a warehouse for storage. Most contractors start using their garage as storage. You may not have a garage, in which case you need a storage space. Consider your storage options close to where you live. You have to be able to store construction equipment, have access early in the morning and late in the afternoon (before and after work), and be easily accessible. Depending on where you live, you can spend anywhere from $50 to $500 a month for a basic 6 foot by 6 foot closet. Space is valuable and you have to stay organized. If you have some cement board saved from an old job, remember it and take it out first before you buy more. You can easily start storing junk and lose track of your space. As with the rest of your business, please be conscious of staying organized. We initially built everything on-site, but now that I have a shop, I can build things easier and better, while using tools and materials not always on the job site. Storage and shop space are options you need to look at as you are growing your business. In the beginning, when it is imperative to start working and building clients, start small and keep expenses to a minimum.

Be mindful of your surroundings. Working in major cities poses many challenges that increase the time and cost of construction immensely. Using tools such as a table saw or a wet saw inside an apartment is not only crowded but poses the risk of damaging interiors. Dealing with condo and coop boards is always frustrating for the homeowner and the contractor, while working by a strict set of rules and working hours. Floors and walls need to be protected in the building common areas daily. The elevator must be used only when scheduled ahead of time for construction purposes. Removing debris is challenging since the garbage must be carried through the building, then loaded into vans that cannot legally be parked outside. Unless there is a major demolition site, you cannot put a dumpster outside the building, and not at a reasonable price anyway. Driving through city traffic delays deliveries and finding parking is a nightmare. Therefore, when pricing your work in a city environment, you have to account for all the timely annoyances and cost mounting intricacies.

DOING GOOD WORK

If you want to achieve success, you must set yourself up for victory. From your organized office to your proper tools, to your exceptional materials, to outstanding workmanship, you have to always be challenging yourself to be better. Never cut corners. Work on someone else's projects as if they were in your own home, and you will never have trouble finding business. Consistently executing all aspects of a project seamlessly will set a standard your business can live by and profit from.

You can continually do good work by properly setting up your business. Use suitable tools and materials.

Hire the right professionals to do specialized work. Look over all the work being done by your crew. Research all aspects of a project beforehand and check that proper techniques are being used. Understand all the pieces of the puzzle and do not lose track of the bigger picture. Look ahead to what is coming and schedule upcoming work accordingly. Understand how much material is being used and why. Always order materials on time so your crew never runs out. Be responsive with your client and communicate with your crew. Always do what you have promised. Respect the time schedule and set expectations for clients. Respect your client and never think they are stupid or will not notice something. Pay attention to details. Be courteous and fair with your crew as you expect them to have your best interest in mind. Cover yourself by putting what you are doing and any requested changes in writing; this will also resonate with your clients and keep them updated.

Doing the best work you possibly can and always striving for better will give you a feeling of pride and invincibility. As the Boston native Ralph Waldo Emerson wrote about self-reliance in the 1800s, "The reward of a thing well done is having done it." You will be respected and you will be in demand for delivering above and beyond your client's expectations. Take pride in what you do and you may conquer the world one project at a time.

Well done is better than well said.
—Benjamin Franklin

Chapter 5
Being the Boss

*If you think it's expensive to hire
a professional to do the job,
wait until you hire an amateur.*
—Red Adair

The day has come for you to hire help. You cannot do it alone. You wish you could have your hands in every part of the project, but you cannot. You are not sure if you can trust others with your company's reputation. You wish you could make copies of yourself, but the best you can do is find someone who shares your work ethic and attention to detail. If you are passionate about the work, you need your crew to be the same. Finding and retaining the right people is essential to developing the business you are aiming for. Just as you couldn't squeeze water from a stone, you shouldn't try to get unqualified workers to deliver exceptional finished jobs. Having the right people by your side is probably the toughest part of your job as a contractor.

MANAGING OTHERS

Managing people could be a book in itself. Corporate managers have bachelor's and master's degrees in the field of management. There is so much to be said about motivating and interacting with other humans. Some people are natural-born leaders while others find it more comforting to follow. Some respond better to criticism and others go further when praised. Some folks are easier to communicate with, whereas you will have a hard time approaching others. Being the boss means getting along with everyone. In the long run, if people do not respect you, they will resent working for you, and your business will certainly suffer. A nice rule of thumb to use when managing others is to treat people the same way you would want to be treated.

I believe in being courteous and giving second chances. Monitoring the work closely is imperative, especially when trying someone new, but giving room for the worker to take responsibility and feel in control is also necessary. Always check the work during the job and at completion to evaluate competence levels. You should do your research and understand how long it should take to complete certain tasks. When watching them work, pay attention to the technique, then give them some breathing room. Some people can stand alone and complete jobs without being monitored, while others take shortcuts and then cover up their tracks if not supervised. Those guys who do not listen and constantly mess up do not belong in your crew. Let them work for the competition. You will find that often-times you are stuck with a guy or two who just does not work well for your team. You may not have the choice of hiring someone else or cannot interrupt the project by bringing in new blood and taking another chance. In such cases, you must monitor them closely and help lead them to the results you need, before the first chance you have to replace them. It is important to always try new guys when available. Taking on an occasional guy for a week will show your crew that you are motivated to finding the best work-force. This will keep them on their toes. You will also see how other people work and possibly learn new tricks, while you may just find your dream worker.

Know your workers and each of their individual strengths and weaknesses. You need to balance the produc-tivity of your team. Ask them what they are good at and what they like to work on, then watch for correlations. It is great if you have the time to teach someone how to work, but offering him or her a free education on your wallet isn't always an option. I have put my tile guy on painting and regretted it when I had to hire someone else to clean up and redo his crooked corner cuts. Learning from mistakes is valuable, but it would be smarter to avoid the mistake in the first place.

Be the leader in your company. You own it, so own up to your title. You need to set standards for your workers to follow. They need to meet those standards and be held accountable. If you tell someone to do something and they do something else because they think their technique is faster, but they delivered poor quality, you end up paying for the mistake. You can only let that happen so many times before you have to start deducting it from their pay. Let your workers know that mistakes happen, as we are all human, but constant neglect will be paid for individually. Two things will happen then: (1) your guys will stop messing up since it is costing them money, and (2) someone who does not listen will just quit since he isn't making much after having to often pay for mistakes.

Changing behavior is difficult, but there are times when you must be hard with your guys and point out mistakes. You should be a coach and not a drill sergeant. Screaming and undermining your people will not earn you respect. They will be even less careful next time as they might just be considering other employers. When confronting someone, it is always a good idea to identify the positives. Let them know that you appreciate good work and point at examples they have accomplished well. Avoid using the word *but*, as this may put them in a defensive mode. Try saying, "Let me work with you on skim-coating this wall so we don't get plaster bumps again." Follow up and keep monitoring the change in that individual. Be helpful to him and inquisitive in his work. The constant reinforcement will solidify the change.

You must be a manager. It is not enough to find people to work, then go practice your golf swing. Managing from far away gives your guys too much room for error. You are not there to provide guidance and answer questions. You will not be able to keep track of the work being done, or lack thereof, and your business will suffer. Micromanaging is also not a good idea. Constantly breathing down your worker's necks makes them feel they are powerless and not trustworthy. Performance will decrease due to this low motivation. You must find a happy medium for a balanced management style. You must provide direction, keep track of progress, and check the work.

There are different working styles necessary for different projects. If you are painting subsidized housing projects, your goal is to work fast and move on even faster. Your margins are probably low, and the expected quality reflects that. If you are renovating a residence in the Plaza Hotel overlooking Central Park, your style and standards of work change radically. The guys you hire on your crew need to work at the appropriate pace. Many contractors stress rushing jobs, patching everything up, making it *look* good, and getting the heck out of there. Workers are used to this style. If you are marketing yourself as a high-end remodeler, you should not be rushing any aspect of the job. You realize how important the details are to your clients and you need to take your time in making everything flawless. Your guys need to understand the type of work expected.

Inspire your workers to be better and work smarter. You are responsible for building the culture and environment you desire. You need to take charge. Your crew is a reflection of yourself. Have them become that. It is a constant trust building effort, but you have to work on the synergies running together for a cohesive team. If you are constantly frustrated that your crew does not deliver your vision, then you have work to do on conveying your thoughts and ensuring they get completed to standard. Construction workers are generally used to working for many different contractors with varying styles. When they work for you, they need to carry out *your* mission and you are responsible for making certain of that.

TALENT SCOUT

Finding the right people to work for you is not easy. Sometimes just finding anyone ready to work when you need them is difficult enough. Depending on where you live and run your business, your pool of laborers may be large or almost non-existent. I am not sure how lucky you may be, but regardless of where you are, here are ways many contractors find workers.

Referrals

The first place to find a worker should be within your network of contacts. As we discussed in an earlier chapter, all of your acquaintances should know by now that you are a professional contractor. Whether you are looking for a helper and a general laborer, or a specific tradesman like a bricklayer, ask around. Does anyone you know have a contact for a person looking for work? Look at friends who have had construction projects done. They might have some contacts. Ask your garbage man. Many times, people working with their hands know others in similar trades, just the way office-type employees have a circle of friends and coworkers with office jobs. Someone knows someone who down the road heard of someone being related to someone looking for work. People love to help each other, especially when in need of work in today's economy, as someday they might need help themselves. You are now an employer, so tap into your network and offer that position to someone in need. This is also where your mentor or other contractor contacts come in handy. I always find that one of my contractor friends has a guy looking for work. When I get the call to recommend a guy, I usually have a list of people I can also recommend. Obviously, once you start developing relationships with workers, they will have other guys they know in similar situations. Start making a roster.

Suppliers

Another place to look for help in finding laborers is with your suppliers. The hardware store, the lumberyard, or the tool rental place almost always have guys who go there looking for work. Go to those places and ask around. You are not selling anyone something they don't want; you are just offering a job. This is not soliciting. Asking if someone needs a job is a noble cause. You will soon make some new friends.

In densely populated areas, the parking lots of Home Depot and Lowe's are full of guys looking for work early in the morning. You will see guys with delivery vans ready to help customers who have no way of taking their large purchases home. The stores do offer delivery service, but these guys will deliver for half the price. Alongside these entrepreneurs are generally laborers waiting to help you install the new chandelier you just bought. These guys are looking for work and you may find the right hand to use that day when you are in a pinch. In some of the New York City boroughs, the pack of available workers is customary. When I go to Massachusetts, however, I find empty quiet parking lots with no helpers in sight. I mention this because if you live near one of these hubs where guys stand in parking lots looking for work, you may be able to tap that resource. Be careful though, some of them may not be legal to work in this country, and local laws may prohibit such shenanigans from happening in commercial parking areas. Legal or otherwise, this is common practice in New York City, and I feel the need to mention it when writing about construction.

Online

Websites like Craigslist offer a place for job seekers to post resumes and for employers to post jobs. First browse through the ads people have put up looking for work. Do a search for things like carpenter, painter, or just construction. You can look through the listings and write down the ones that seem promising. Call or email them first before you post an ad offering a position. It is easier to interview the people you want than to get inundated with calls from folks whose skills sometimes don't even match your described requirements. Sometimes you will call on one of the guys who has an ad looking for work. It turns out he is not a laborer but more of a handyman who runs his own jobs. He won't come work for you at a daily rate, but he does know a few guys he uses who are also looking for work. And so the networking continues. If you do decide to post an ad offering a job, be ready to interview candidates. Be honest with your expectations and do not lead them on. If you only need someone for a weeklong project, state that right away. You would not want a worker to tell you he knows how to tile when in fact he's only a painter, and you do not want to give him false hope either. Be sure to ask about experience, tools, and transportation means. If the phone interview goes well, you should meet in person, on the job site preferably, to finalize the hiring process.

Subcontractors

Sometimes you just cannot find a guy to do the work you need for a daily wage. That is one time when you need to just hire a subcontractor to get the job done. If you bid that part of the project correctly, maybe you can still make a little profit. If not, the work still needs to be finished. Be honest with your sub and let them know you are losing money. They sometimes feel your pain and work with you at a lower rate. Sometimes you have a very specialized part of a larger project that you and your guys cannot complete since it is way out of your expertise area. I subcontract all my glass and mirror work. I have no desire to acquire glass cutting and shaping tools, nor do I have the space to store and work with large sheets of glass. This job is best left to the professionals, on the occasions I need them. A third time you may need a subcontractor is when you are mandated by building rules. As an example, in New York City, most apartment buildings require a master plumber and master electrician to handle any plumbing or electrical work. While my crew has the expertise to properly finish most plumbing and electrical jobs, and while I am confident we would meet local code standards, the buildings in which we work will just not allow it. They are covering themselves from major issues that could arise from using unqualified work. Therefore, when the building permit requires you to use licensed professionals, you have no choice but to subcontract that work.

You should start a list of subcontractors you have used and ones who may be of use in the future. Having this directory will make your life so much easier when pricing jobs and needing to piece components together. Many times I have priced plumbing and electrical work based on bids from subcontractors, knowing that would be my maximum cost. If I was in a squeeze and had to use them, I would be covered. If I had the time and authority to do the work myself, I would be making more money. Covering yourself for the worst-case scenario is smart business. Just as above, subs can be found from referrals, suppliers, and online. One other place to find reliable subcontractors is a simple Google search. There are many online listing services that will provide you with legitimate companies. The same websites you may use to list your services might be marketing your subs as well. Search for what you need, and call a few to get quotes. You will start establishing relationships and understand their pricing as time goes on. Having licensed professionals that are by your side and ready to work with you is invaluable in your business roster. Subs can sometimes refer other subs or individual workers, and may be proficient in other areas you could use on a project. As mentioned earlier in the book, subs can also teach you things about their trade or your own if you are open to it.

A cautionary note on subcontractors: Be careful! When taking bids for a specific job, like a plumbing portion of a larger project, make sure you are comparing apples to apples. You will inevitably get price quotes that seem exaggerated and some that seem too low. Make sure you interview your subs and proceed with caution. The cheapest price doesn't always turn out so cheap. Some subs will low-ball you to get their hands on your project. When the time comes to complete the work, they will keep adding on cost, as apparently all that work wasn't included in their original price. Some subs will send apprentice workers to do the job. You will not only get inferior work, but they will make mistakes and tie up your time having to come back several times to fix their own problems. Before you hire a sub, ask specifically who will be doing the actual work and go over *all* the work you need. Only then can you correctly compare price and quality. Also, hold a larger percentage of their pay until they finish their duties, especially when you do not have an established relationship.

KEEPING TALENT

Regardless of how you find your workers, some may end up working for you for a long time, some may get occasional callbacks, while most will just remain nameless phone numbers in your cell phone. Some of your temporary laborers will organically stay working for you, and just naturally become a part of your permanent workforce. There are times when you will have to make formal full-time employment offers in order to recruit the best guys.

The most skilled construction hands generally do not have to look for work. Work will find them. When they are talented, driven, and detail oriented, their employer will want to keep them. They can think on their own, follow direction well, measure twice and cut once, fix mistakes and solve problems, take pride in their work, and they are hard working and always stay busy. They are making their contractor money and saving him from aggravation, so he will want to reward that if he is smart. Those are the guys all contractors want. Those are the guys you need. They are the foreman, superintendent, and lead carpenter in one body. They are a reflection of your ambitions, minus your business savvy side. How do you get the elite working for you? You have three options:

1. You could strike gold and randomly find one
2. Take on a young guy with potential and help him grow into your ace
3. Steal him from your competition!

You always have to be on the lookout for potential talent that can strengthen your business. Some contractors do not offer full-time work, so in between jobs their crew is looking for work. You may also get lucky and grab one excellent guy while he is trying to get some side jobs to make extra money for Christmas. Sometimes, when you see a guy working elsewhere that would be a great fit on your team, you just have to make him a better offer. In my own experience, I have found that if someone is willing to consider leaving their employer to come work for you, they weren't quite so happy there in the first place. It is not unethical to hire away a worker from another contractor. It is just business, and business creates competition, which keeps everyone on their toes. You can win contracts by doing work either faster, or cheaper, or better. All three ways involve having a competent crew. I have had guys politely refuse my offer, or several offers, and I have had guys who gladly seized the opportunity. Regardless of how you get your hands on a diamond, be smart and hold onto it. There is always the fear of not having constant work and losing your best guys. This is why you have good guys, to let them work while you go and book the next job. You are now a business owner and need to make your business succeed. So more often than not, you should drop your tool belt and go enamor some clients. Of course, that is if you want your business to grow and not just do one small job at a time. You are in charge. Making the commitment to hang on to your best performers does not have to be so daunting. It is worth having a guy you can depend on, even if you are paying him while there is no work. He has *your* best interest in mind, and you should have *his* as well. At times when work is slow, have your full-time guy come to the shop (or whatever setup you have in your garage or basement) and do maintenance work for the business. Clean tools, sharpen blades, organize materials, build jigs, research construction techniques for upcoming projects, or play ball with your kids while you're busy marketing. Regardless of what he is doing, he is an asset to you and you need to keep him busy. He will be right there for you when that small project pops up or for that big renovation just around the corner. At times, you can let them go find side jobs if you are really slow, but just remember to have some sort of incentive in place for them to come back to you. It is wise to account for their salary as a fixed expense for your business. This way, rain or shine, he is a part of your business, just the way insurance and rent are. Having these fixed costs will make you work that much harder at finding and booking jobs. Just be very careful not to keep on your payroll the guys who cannot work without supervision just because you feel like you must have someone available to work. There are always more run-of-the-mill personnel if you just look at another mill.

A great way to keep your employees happy while at the same time motivating them to work better is to incite good behavior. Giving them a bonus for exceptional work will keep them working for you and will make them strive for better to get yet another bonus. Financial incentives, time flexibility, even buying lunch or having shorter summer hours can all be positives for your crew. For example, if you bid on a job with an estimated eight-week completion, tell your crew they will get two free paid days off if they finish a week early. Make sure their standards do not drop, while the job gets rushed at the expense of quality. The characteristics of the finished product are always the priority. Workers can sometimes be careless and waste materials on jobs. They leave paint at the bottom of buckets, they mix too much plaster that ends up in the garbage, they leave the glue cap off which dries overnight, and so on. Pay out a percentage of the savings in materials cost if they use it more efficiently and come in less than projected. Again, be watchful; using less thin-set to lay tile is an easy way for them to get a bonus but a hard way for you to get a callback. Also, make sure the value of whatever bonus you are offering is at least covered by the savings in your projected bid. If you are paying a guy a bonus worth two days of work, but you are saving one day of labor, your profit just got slashed by the cost of one labor day. It doesn't make much sense now, does it?

HIDDEN COSTS OF LABOR

Unions have a bad reputation these days, or at least through the eyes of non-union folks. (I am not referring to any of the public sector labor unions for jobs like teachers and police, though many controversies stem from those as well.) Unions were one of the major causes of large-scale bankruptcies like General Motors, and have continued to make projects cost more and take longer to complete. The well-being and financial gain of the union workers and their invincible attitude antics are another story. We have all seen the giant rat at the site of non-union jobs. The truth is, we are going to be seeing more and more rats. The percentage of the labor force belonging to a union peaked in the 1950s at around 35 percent, and sits at less than 10 percent today. Fewer employers want to deal with unions for many reasons not worth getting into here. To the union workers' justification however, they are generally dependable and produce consistent high quality work. That being said, union or non-union, superior workmanship will always cost more.

You may think you are getting a bargain for hiring a guy at half the going rate for a carpenter. He seems to know how to do the work and can finish tasks. He does make a few mistakes here and there, but he can correct them. This seems like a winning situation to you. Your labor is way cheaper and you're finishing jobs just fine. Be careful now. This guy is a helper and he is getting paid the wage of a helper. He should not be a worker in charge of completing duties on his own. He is not quite there yet, and his capacity may never allow him to get there. How fast is he finishing his tasks? How many mistakes does he make? How much time is he taking to fix the mistakes? Who is paying for all this wasted time? How many pieces did he cut wrong that are now wasted material? Who is paying for the scratched customer's furniture he just nicked while being careless in their home? All of those mistakes add up. If you are paying an amateur twice as little but he finishes his tasks in half the time, you are losing money by not hiring a pro. On the surface, spending less on labor seems like a good idea, but in the end, the quality of labor more than makes up for your cost. You may still not believe me if I give you theoretical examples. You have to see it for yourself. I promise you though, the aggravation, wasted materials, time exhausted, and likelihood that your clients will notice the amateur work, will keep your business from reaching its full potential. Many contractors are stuck in an endless cycle of fixing mistakes and taking too long to carry out inferior work. Build your business precisely to become an elite contractor. You will save yourself more than just time and money, you will become more profitable, and you will feel more confident in your abilities to deliver on promises.

In your career as a contractor, you will undoubtedly run into unreliable workers. Be on the lookout and be careful while monitoring your guys closely, especially in the beginning. On top of the fact that you will run into sloppy guys who just cannot seem to be able to keep a clean work environment or finish something without damaging something else, you will notice behavioral problems. Some guys like to party too much. What they do in their own spare time should not matter if their work is stellar. But oftentimes, problems outside of work will interfere with work. You will have a carpenter who works great with you for a few weeks, and one day he just doesn't show up for work. As chance would have it, that would also be a very important day as you plan to finish a section of your project for the customer to review. You try calling repeatedly but his phone is off. Is he all right? You may find out when he decides to come back to work, or you may simply never hear from him again. I cannot explain such behavior, but I can tell you that it happens. Not everyone is as responsible as you may be. I have had guys show up for work reeking of gin. I have found empty bottles of Corona in the construction garbage. I have seen guys at work acting like they are high on something that does not resemble caffeine and were laughing too much to have been drunk. Sadly enough, there are plenty of people out there who have drug and alcohol issues. Just be mindful and watchful for such behavior on your job sites. You may also hire guys who disappear for the simple reason they got a better offer elsewhere. They will not give you notice or a chance to counteroffer. They just won't show up. Being the boss, you should set the tone. Promote honesty and communication within your crew. If you are approachable and easy to talk to, guys may have more respect and give you more credit. Allow room for guys to be open and share what is bothering them. Let them know it is better to leave the door open than to burn the bridge. Simply put, if you have a problem and cannot come to work, give me notice, even if that same day. It gives me a chance to hire someone else or try to fix the problem, and I may use you again for being truthful and respectful toward my business.

There cannot be a book written about construction that does not address illegal aliens working in the field. Admit it or ignore it, fight it or embrace it, love it or hate it, participate in it or report it, undocumented workers are a major part of our economy. Whether we like it or not, laborers who are not legal to work in this country are used for construction every day. Debating immigration laws is a whole other issue better left for Capitol Hill. This country was built on the basis of hardworking families who migrated here in order to better their lives. At our roots, we all started elsewhere and ended up struggling to build this great nation. Indeed, we are privileged to live here, a dreamland built by our colonist ancestors. People feel very strongly about immigration, so I am not here to dispute it but only to present it to you. It is illegal to hire someone not authorized to work in this country. Generally, undocumented workers are willing to accept lower wages. Due to this labor cost savings, contractors hire illegal help on a regular basis on construction jobs. You may be competing for projects with someone who will always underbid you due to their lower overhead. Since you cannot lower your prices any further to profitably stay in business, you decide to hire cheaper labor. This means either hiring guys who have no skill and will ruin your work, or hiring some guys off the books. This is a perpetual revolving door. In smaller towns across America, this may not be as prevalent of an issue, but in large cities where immigration is in full force, the cheap illegal workers are easily accessible and used.

Many individuals without the ability to get a Social Security Number are allowed to have an Individual Tax Identification Number. This number allows them to pay taxes on money earned, gives them access to a bank account, and even state IDs and driver's licenses. On paper, they will blend in with the rest of the workforce. When you are hiring independent contractors and providing them with 1099s for tax purposes, these illegal workers can easily fill out the same W9 form as any other worker. They will work and pay taxes just as any other subcontractor would. Many times, contractors may be using employees not allowed to work in this country without even knowing it, since you would not need a Social Security Card for an independent contractor. Independent contractors are individuals or businesses that work for you but are not on your payroll. They are self-employed and are not your employees. You hire them to provide a service to your business. These workers pay their own taxes and you do not withhold taxes from their checks. Though sometimes misclassified, for tax purposes, independent contractors are very common in the construction world.

The reality is that undocumented workers are often used in the construction field. Whether you agree with it or not, many illegal hands have built and renovated our homes. Most likely, if they did not exist, there would be a much smaller pool of construction workers with higher wages. You may run across this issue in your career. As you will see, many contractors make their own decisions in hiring illegal work. Whether you think it is right or wrong, how you interpret and enforce the law is up to you. Just understand all consequences of your actions before you make any lasting decisions.

So what should you expect to pay workers? That question will have varying answers depending on where you live, the demand for that specific job, and how many tradesmen are available in that field; a simple supply and demand function with a location variable. You will be able to answer this question on your own based on feedback from other professionals in your area. I will give you a range you should expect. In construction, day laborers make between $15–$25 per hour, skilled tradesmen make $25–$50 per hour, while union workers make above $50 per hour with benefits packages scoring them wages of over $100 an hour. Yes, you read that right! If they were to work full time at that rate, their salaries reach a quarter million dollars per year. Don't run yet! There is no need to quit your new business and join a union. While their pay is fantastic and their benefits give them a golden retirement, many of my friends in the union sit without work most of the year.

I would hire them, but if they work outside the union, they get thrown out and lose all their benefits. While you may be on a cushy city project for eight months, you could be spending the next two years sitting at home watching the kids. I have a union electrician friend who I would consider a nanny by trade and an electrician by hobby.

In the major cities, you will be paying most of your workers between $150 and $200 per day. In less dense areas, your labor cost can be up to 50 percent less. When you hire a plumber or an electrician, these qualify as subcontractors and will not work for you at a daily rate. They will generally give you a price for the job, regardless of how long it takes them. They usually do not make less than $300 a day and they will not show up for jobs smaller than a day's pay, unless you have that kind of relationship with them.

The daily rate you pay your workers is not your only labor expense. If you pay them as independent contractors, as you will for many of your guys, you only need to declare the income, as they handle their own taxes. But if they are employees, you have to factor in several other costs. Employer taxes include social security, medicare, disability and unemployment insurance, and possibly other taxes depending on your state. The federal payroll tax you will be responsible for for each employee is currently 7.65 percent. State payroll tax will vary between 2 percent to 10 percent depending on your state. Add to the tax your cost for worker's comp for each individual employee, and you will see your cost of labor rise another 20 percent or more. Depending on your state, the amount of time you have, and the employee, you may be required to provide other benefits such as health insurance. Employee benefits can add up quickly and you may be looking at skyrocketing cost of labor. In conclusion, when you hire someone to work for you full time, think long and hard about that person and your added cost. You will need to price your jobs accordingly and factor in your true cost of labor, not just the guy's daily rate.

Your employees will be your extended family. At times, you will actually be spending more time with them than your friends or real family. Never forget to enjoy what you do. You are in this business for your own personal reasons, but I hope being miserable is not one of them. Enjoying your work and being brilliant at it are that much easier when you have the right people around you. Putting together a dream team that creates a safe, productive, and enjoyable atmosphere is difficult. You will have to work tirelessly at it, understand your employees' strengths and weaknesses, and share their values. You will also need to set expectations and goals for yourself and for your crew. Know what your team is worth and charge accordingly. Most importantly, if you cannot be proud of your team and their accomplishments, you are marching in enemy territory with the wrong troops.

Chapter 6
The Client

Nobody cares how much you know,
until they know how much you care.
—Theodore Roosevelt

I love building things. I always have. I get a high out of seeing projects flourish. It is in my nature. If I have free time, I am always building or fixing something around the house. As a kid, I would constantly be *fixing* clocks and radios around the house, none of which ever needed fixing. I would actually take apart anything I could get my hands on. Little did anyone know that it was practice for what I would ultimately turn into a business. A passion for constructing is wonderful, but how far would it get me if I had no customers to build for? DIY projects around my house certainly do not pay the bills. I take my client relationships very seriously in order to keep my passion for construction flourishing. Customers are the ones who keep me in the business of building things. My clients afford me the privilege of continuing to be a kid.

The synergy between my passion and my clients is artfully summarized by Ayn Rand in her novel about an architect, *The Fountainhead*: "I don't build in order to have clients. I have clients in order to build."

BOOKING THE JOB

Historically, the bar has been set fairly low for customer service in the construction industry. People are used to contractors being tradesmen and not much else. With a little care and attention to the client, you should stand out above most other contractors. Homeowners care about being attended to and want to ensure their goals are met. Catering to their needs is particularly important when in the bidding phase. You want your prospect clients to trust you in their home.

First impressions are lasting. When meeting someone for the first time, it doesn't take you long to judge whether they are responsible or to see if you want to learn more about them. Would you want to be their friend? How about working with them? You can generally tell in the first few minutes of a conversation if you would be able to get along. As a freshly minted contractor, how would you want a client to judge you? Here are a few considerations that will set you up for success and on a path to signing your next contract:

Look presentable on paper

A client's first exposure to you is most likely virtual. If you get your leads online, your website will be their first foray into your dealings. Maybe they were a referral and you first introduce yourself through email. Step back and consider how you would want to be approached by a contractor. Maybe ask a friend to review your website or an initial introductory email that you send out to leads. Before meeting the client in person you will have to speak on the phone for an introduction, an overview of the work needed, and to set up the meeting. You should sound enthusiastic about the project and you should be available to meet as soon as they can. You may be asked if you are busy with work at that time. Keep in mind that most successful contractors are always busy.

Look presentable in person

Showing up with sawdust in your hair might not even get you inside the home. Coming to the meeting in a suit and tie might raise questions, like if you understand actual construction work since you've probably never used a hammer in your life, or if you are too expensive, given your look. People want to hire a hardworking and reliable contractor. Be clean. Always wear a clean shirt, pants and shoes. Leave a change of clothes in your truck if you must go from a job site. At the very least, wash your face and make sure you are not dusty. Smelling bad is unacceptable, though smelling like bathroom cologne will also put most people off. Your appearance says a lot about how your job site will look. Take your shoes off. Always, without exceptions. If the homeowner tells you otherwise *and* you see they are wearing shoes, then it is ok to keep yours on. Without a doubt in your mind, when you walk in the front door, take the shoes off. If the clients are being polite and tell you to keep them on *but* they are only wearing socks, it is *not* ok to keep yours on. People want to make sure their home stays clean before, during, and after construction. Never bring food or drinks with you. Leave your jewelry at home and wear it off the job; you are not at a fashion show. Be mindful of your tattoos. If you have tattoos visible on your body, you should probably cover them up for your first meeting. Maybe wear long pants and long-sleeved shirts. Err on the side of conservative and look presentable to any gender, age group, or cultural background. Take your hat and sunglasses off when speaking to your audience. You want to look approachable, not like a standoffish teenager.

Be respectful

Be aware of people's personal space. Never set your tape measure or anything you may have brought to the meeting on the customer's furniture. Being considerate of belongings in the home should also be a priority for you and your crew when on the job. Showing that you are caring during your first meeting will put your client at ease with having you in their home. You are essentially invading their most private space and they need to trust you. Never use curse words; clean up your language and think before you speak. Do not say any jokes that may be offensive to some. Never touch the person with whom you are speaking in any way. Tapping them to show them something or holding their shoulder while laughing is not acceptable for a person you just met. Always be professional; you are meeting a client, not an old college buddy. Any questionable actions should be left for your poker club night. Speaking negatively about someone else, or debating religion and politics is not advised. Always be courteous. You are winning this game if you can speak and act as you would in front of your old English teacher.

Show interest

Do not check your watch, phone, or the TV for the score of the World Series game. The client's construction problems are your number-one priority at this point; everything else can wait. Listen. Let the client speak and present the project. Do not finish their sentence or interrupt them. Again, listen to them. While you are listening to what they have to say and watching what they are showing you, formulate answers in your head. They will ask for your advice on certain things and you should have solutions on the spot. You must seem knowledgeable in all areas. From your first phone conversation with them, you should have a general idea of the scope of work. Before you get to their home, do a little research on that type of work. You must be educated on the subject and be ready to offer ideas. If, for example, they want their floors refinished, do not assume you will get a subcontractor to do the work and you will just get a bid without talking to your clients about it. They will ask you how long it takes, how much dust it makes, how to get a darker stain, if the gaps in-between the planks can be filled, and God knows what else. You must have answers and provide options on the spot. Know what you are talking about and do not make stuff up. These days, I find some clients research their construction needs before they call me. They are educated on what needs to be done and have read reviews online from other people in their shoes. Be interested in their project and have some of your work to present as examples. "Oh, you need your ceiling fan moved centered with your bed? Perfect idea; we just did that for another client and she loves it. This is how we did it." At this point you can up-sell by recommending that a dimmer switch be installed or possibly other electrical work you see around the house.

Follow up

Always be available and responsive with your client. If you get an email, answer it right away. If you are busy on the job or you do not know the answer, reply with a quick "I'll get back to you before the day's end," then set yourself a reminder to research the topic and answer the email when you get home. Return all phone calls or text messages within thirty minutes. For a prospective client, do not give them a chance to call the other contractors bidding on the job. When they call, drop the tools and pick up the phone. Do not give up on a prospect customer. Do not assume they lost interest if they do not follow up. You should follow up. Respectfully stay on their case and show interest. They are busy with their work and lives so you need to remind them how ready you are to make their dream project a reality. Renovations are a big step for homeowners and they often need time to get used to the idea and to digest your proposal. Check in periodically until you are specifically told to stop calling because they booked another contractor. At that point, you have an opportunity to learn from your mistakes by asking for some honest feedback on why you lost the bid. Generally, most of your competing contractors give up on that customer who is a tough sell. You being the last one standing have better odds, as they will remember your persistence and how responsive you are to their needs. Show your clients that you want this job. If you seem too busy to attend to their needs, they will assume you will not focus on their projects and will want someone else who will make them a priority.

Be nice

Always be positive and optimistic when dealing with clients. You will get better responses if you always smile and show you are a happy person. Your image should be an eager, good-intentioned, ready-to-help person, not a grumpy service provider who only cares about his profits. Do not assume people are stupid. Speak in plain language but do not underestimate or undermine your clients. Leave your drama and personal issues at home. Other people have their own; I assure you they are not interested in hearing about yours. Plus, you will look too fixated on your own issues to properly address their renovation concerns.

We spoke earlier about always giving prospective customers relatable examples of your work. Make them realize you are the perfect contractor for their job. Let them know you have successfully completed similar jobs. It will put them at ease knowing you won't be using their home for practice. Also, as you look at their particular project, show them you are informed and know a multitude of areas. When you walk through someone's house, you will most likely notice plenty of ways to improve it and add to their construction list. In newer apartment buildings, as they were built to code, you will find things where they should be. In dethatched homes, you may run into an array of non-standard procedures done and redone by handymen over the years. By pointing out code violations and why they are also unsafe for the homeowner, you may get yourself more work in the future. *I am here to stain your deck but noticed your outdoor receptacles are not GFCI protected. The code says . . . and it is safer for you because . . .*

As mentioned earlier in the book, doing market research is very important to the sustainability and development of your business. You must figure out what your clients want and provide that service. You must also constantly seek improvement and monitor to make sure your clients' expectations are met. You always should be accessible and responsive. Communicating with your client will avoid costly mistakes and having to redo work later. At the end of the project you need to ask your client for honest feedback. What things were you good at and where do they see the need for improvement? This will not only help you get better, but also show them you care about the customer.

Ask your workers the same questions. Always look for strengthening efficiencies within your company and ways to enhance the customer experience. The moment of truth comes when you ask the question "Would you use our services again or refer us to a friend?"

THE CLIENT IS ALWAYS RIGHT

You have heard it many times before: *the client is always right*. Now that you are in business for yourself, it is time to start living that notion. Even when the client is wrong in your eyes, they are still right in their eyes. Their point of view is really all that matters, because they are the ones writing the checks. They keep you in business, so their perspective is the right one.

How do you deal with lousy customers and problems that inevitably arise? First and foremost, always do a good job! Adhere to the highest standards and instill that notion in your crew. You want to do the best work you possibly can, and then figure out how you can do it better. Being on top of your game is essential for two reasons:

- Your clients will kill your business if they are unhappy
- Your own pride, moral standards, and integrity

First, you cannot afford unhappy customers. Remember Warren Buffet's quote about your reputation? One dismal client can ruin you online. With so many people checking reviews online these days, your welfare depends on being portrayed in a positive manner. The world isn't always fair, and there are times when you know you are right but you still have to please the customer who is wrong. Customers will sometimes try to take advantage of you and situations that make them feel they have leverage. There is no reason for you to let a conflict escalate to such proportions that a customer feels compelled to trash you in a review. You simply cannot afford to be dishonored online or in your community. This is why you should always remember to build in an error margin in your bids. You will make mistakes and you need to have a financial cushion to handle them. Please the customer even when they are wrong, and make them feel respected. I remember a motto used by traders in the financial markets that holds very true in this scenario: *Live to fight another day.* If you choose to fight your customer, you may win that battle, but you will lose the war that awaits you out there in the construction world. Give in and always please your customer. The root of all the problems is something you as contractor did. No one would ever have an issue if everything was done right. The truth is that maybe your customer is blowing it out of proportion and adding unrelated things, but it was all stemmed from your mistake. Somewhere down the line you dropped the ball. Take full reasonability for the fact that you are working in *their* home and something *you* did lead to an issue. Fix it. Pick up the ball and don't drop it again. If you do get a negative review, respond to it. Write your comments right next to it for everyone else to see. Write your specific actions on how you offered solutions, how you fixed the problem, and how you presented further support. Prospect clients reading the negative review need to see you are open to clear up any situation. We are only human and make mistakes. How we deal with challenging situations is what differentiates us.

Second: Be ethical. Doing the right things should not even be a debate or ever a question in your mind. Cutting corners and installing makeshift creations inside the walls and only covering them up later to make it look presentable is offensive. If you do not know how to do it, look it up or ask someone before you make it up. If you are clueless, do not assume your guys are experts. Check the work and ask questions. Be accurate when being entrusted with other people's safety and allowed to work on their homes. If you see a problem or spot a mistake, fix it before it ever becomes an issue. Your clients should never see a mistake, and you should never cover it up. Deal with it before it gets out of hand. Let your crew know that covering up mistakes is unacceptable on your jobs. Built quality is more important than time constraints, as we take the time to go back and fix errors. This small time cost can get amplified later and can turn into perpetual carelessness. Build your projects for posterity. Construct everything well built and be confident they will weather any storm. Not only will you avoid callbacks and disappointed clients, but you will know in your heart how your work will stand the test of time. Your mark should be of quality and vigor. A rushing superficial contractor is not memorable. Have pride in all that you do. To most contractors out there who choose to be sloppy, I thank you. You are creating more work for me. I have gotten more work because of your substandard client satisfaction, choppy and inferior work that I need to repair, and I have won bids over you because of your questionable etiquette. If you are reading this book, I am positive you will not become a careless contractor. Without getting too spiritual, consider karma in life. Do good and good will come back to you; be ruthless and I am uncertain of your future.

Chapter 7
Contracts

By failing to prepare, you are preparing to fail.
—**Benjamin Franklin**

You meet a new client. You hit it off right away. They are so easygoing. You present them with exactly what they were looking for. The price fits their budget perfectly. You think this job was heaven-sent. They will be your customers forever. Nothing can possibly go wrong. There is no pressure on the timeline and they seem to trust you completely. They give you a deposit. You shake hands and get to work on what you verbally agreed upon. From then on, Murphy's Law takes effect. Good luck to you; you will need it.

I have made agreements where I felt completely comfortable with homeowners, designers, subcontractors, and even friends. Somehow, almost always, I seem to get the short end of the stick in these situations where there is no written proof. Not only have I lost money, but wasted time and dealt with unnecessary frustration, all of which could have been avoided with the proper paper trail. I hope you can learn from my mistakes and never start a business endeavor without a written agreement. Something always seems to go wrong and the words exchanged in the beginning seem to be forgotten. Solidify your job and protect yourself by always having a signed contract or another form of a written acknowledgment.

THE IMPORTANCE OF CONTRACTS

Contracts are not just a legal bother to be used just in case something goes wrong. They are also clear reminders to both you and your customer of what needs to be done. Many times, people see things from their own perspectives. We all sometimes only hear what we want to hear and neglect other facts. Two people can leave with two very different sides of the same story. Written agreements give you the opportunity to deal with issues before the project ever starts, and avoid misunderstandings.

Outlining everything beforehand and having it on paper may deter the unwilling or unable of paying customers. Sometimes folks aspire too much without having the resources to back up their ambitions. If they cannot afford the renovation, they may think twice when they have to sign next to that hefty price. You may be presented with the option of cutting out some of the work before you complete it and wait for payment for years to come. I hope you never have to work for a client paying you in IOUs, but having a contract will certainly aid in preventing that.

If you find yourself in a tight spot, you do have rights you can fall back on. Having a signed contract and other hard proof of agreements gives you a legal recourse. Though you should certainly try to avoid legal action at all cost, having that backing gives you leverage.

As we spoke about your selling image earlier, being seen as legitimate should be high on your list as a business owner. Preparing a client for what goes in a contract and presenting formal documentation will confirm your company as a solid entity and you as a trustworthy individual. The client will take you more seriously and treat you as the professional you seek to be. You perform quality work and have high standards, so present yourself in that light regardless of how large or small the project is. Showing my professionalism in presenting thoughtful proposals has gotten me larger jobs from small handyman-type of work. People will trust you further once your commitment is apparent, on paper or in person.

TYPES OF CONTRACTS

Many states and condo associations require a signed contract between the client and the contractor before any construction work can take place. This protects the consumer, the service provider, and makes the work legitimate and trackable. You should check with your local government for specifics on what the law requires. Consumer protection agencies also require certain things to be included in a contract. Make sure your contract is legal and includes all required details for the area in which you are working.

AIA (The American Institute of Architects) provides standard contracts often used in construction. Major work and government projects are usually contracted on lengthy and intricate standard contracts. The issue for the small contractor, with these types of documents coming from an architect or a GC, is that they first protect the architect, then the consumer, and lean heavily against the rights of the builder. They may not be fit for you without alterations and additions. Please *never* sign a document you do not fully understand. The reward is never worth your risk. Always take your time understanding a contract and seek advice if still uncertain.

A contract, however, does not have to be this obscure manuscript. It can simply be a few paragraphs in which you describe what is expected of you and your client. Most of the time, especially in the beginning of your career and when handling small jobs, this is all you will need. You will write it yourself in clearly understandable English. There is no need to copy and paste legal verbiage. Expectations should be easy to understand and agree on, without confusing anyone with official terms and old English pronouns.

The following is easy for a customer to sign promptly and will protect you in case the client decides to change their mind on payment or conveniently forgets what is expected:

Joe the Carpenter King   *ATTN:Jane Customer*
100 Trust Lane   *200 Safe Road*
Happy Place, VI 00804   *Happy Place, VI 00804*

On Monday, June 21, 2015, I will remove your old base-boards in the master bedroom and replace them with new ones that you are providing. I will bring tools and other necessary construction materials.

The room will be completely empty of any furniture or belongings and your children and pets will stay out of it while I work. I am not responsible for anything left in the room.

I will provide standard quality of workmanship and I will be paid $250 at completion, that same day–06/21/2015.

With kind regards,
Joe King   *Customer:*_____
06/14/2015   *Date:*_____

Now you've got a signed contract. Go change those base-boards with a smile!

You will be signing contracts with various parties depending on the job. You may be the sub on a larger project where the GC contracts you. You may hire your own subs with contracts to fulfill their part. Know all the nuances and details of the work to come. You are required to do that work if listed in the contract. If it is not written, you are not responsible for it.

A contract may also be just a simple email. Anything that can be tracked qualifies as a contract. If you asked in an email if it is ok to proceed with a certain wall color and you got an affirmative reply, you have saved proof that cannot easily be contested. You also have a reminder for your client to see if/when they try to argue with you about how they wanted the blue they had in their head and not the red they agreed to in writing. Always, always, always, get confirmation for anything in question with a simple email. Take a picture of what needs to be verified and send that to the client. Ask for approval. You need to cover yourself from issues that may arise from absent-minded clients. Anything you do in a person's home needs to be verified by them beforehand. Emails with pictures are a quick way to get approval without formalities. Never take someone's verbal confirmation as proof. People conveniently turn forgetful at the opportune time.

Most contracts are made on the basis of a lump sum, but there is yet another type of contract you may consider: time and material. When first starting out, you may be tempted to work for time and material with a markup, instead of risking pricing a lump sum incorrectly. Doing such work can also provide a track record for you to price future jobs. Be careful of time and material contracts, as customers get suspicious and micromanage your projects when they know the cost falls on them. I have agreed to work for time and materials for friends before, to try to minimize their cost, but at the same time I have shortchanged myself by limiting my ability to make a profit. I try not to make such contracts unless it is for a change order that may get out of hand with unknowns. If you agree to work on time and materials, specify to your client that you will track all the labor, including a salary for yourself, all the materials, and add a markup of around 20 percent for overhead. Make sure to include any overhead you can directly attribute or proportion to that job, charging the client directly for it. New York City, for example, limits the amount contractors can mark up to 21 percent. Often, for private jobs, the markups can be 50 percent to 100 percent. As you can see, your goal should be to price your jobs correctly and charge a lump sum to give yourself ample room to make a profit.

PARTS OF A CONTRACT

Here are a few issues you need to address beforehand in a contract. These are things important to consider, some of which I have learned the hard way. Consider the following carefully and protect yourself before something happens. Addressing these points may just stop them from happening in the first place.

You will often reference other documents in your contract. Your estimated bid, architect's drawings, and other specification sheets become a part of your contract when you note them. Instead of writing your entire proposal over again in the contract, just include a note indicating the work as described in the proposal is to be done accordingly. Be careful of larger projects where you are required to complete certain work by the architect. You may be agreeing to complete more than what you had expected. So be specific when you reference other documents.

Date your outgoing documents. You need to always date your estimates and contracts. Anything in writing from your company needs to be dated. You should also provide a timeline for which your proposal is valid. You do not want someone to see a price and try to hold you to it for an indefinite amount of time. By specifying the price being valid for let's say thirty days only, you give yourself time to adjust it and provide a sense of urgency. You may be busy after that time or supply prices could have gone up. Leave room to charge accordingly when your bid expires.

Be specific on what is included and what is not:

Next to your total price, write if you are including sales tax or not. Do not shortchange yourself of up to 10 percent of the project cost that the state will be collecting from you. (Tennessee for example has a 9.45 percent state and local sales tax.) Your price is your price; the consumer must pay the sales tax; make it clear it is not your responsibility, but your legal requirement to collect it and send it to the state.

Obtaining work permits may or may not be included in your cost and time frame. Clearly address that in your contract. Most often, the contractors obtain the authorizations for work and the permits required, because we understand them and deal with them often. They require time and money. Do not neglect them and assume you can just work without filing any paperwork with local government agencies or condo associations. Your client may say you are all set to start work. Do not do so until you see approvals in writing. Working without permits will get your jobs shut down, your business fined and monitored, and your license in jeopardy. The contract should state who is responsible for obtaining them.

Your quoted price does not include fixtures. All items to be installed such as light fixtures, faucets, cabinets, stone counters, and other variable cost items are the responsibility of the client. Fixtures of unknown styles vary widely in price. You need to be specific in your contract about them being the owner's responsibility. In the rare occasion you are asked to purchase them, you need to specify in writing the understanding that you will bill such costs to the client. Always be explicit in a contract about what exactly is included in your price and what is extra. Construction materials you can estimate, but fixtures the client would want you cannot.

Labor, construction materials, and tools are usually all included and should be noted in the contract. There may be special tools you need to rent in certain circumstances, or the client may be providing certain materials. Cover them in the contract.

The timeline should be clearly identified. Make sure your time calculations are accurate, then give yourself some cushion and say the project should take slightly longer. Write the exact start date and the end date of the job. Make sure you specify that delays out of your control can change the finish date. You should not be held responsible if you do not start on time because the client did not get you keys, or did not finish clearing the job site, or never got the proper permits for work. Some formal contracts provided by GCs or architects have financial penalties for not finishing on time as well as incentives for finishing early. Delays beyond your control may include the loss of key staff or subcontractors. You need time to replace that workforce and you should set that expectation. Weather conditions need no explanation, but they do need to be mentioned in the contract. If the client postponed the job, it should be on his dime. You have a crew to pay and cannot afford to wait around a few more weeks for your client to go on vacation before the job starts. You will start work on the date agreed, otherwise the client pays for your lost time. Also, rescheduling work during the course of the job comes at a cost. Mention time restrictions by the condo building if applicable. I have had to work on apartment renovations in buildings where work is permitted only between the hours of 9:30 a.m. and 3:30 p.m., Monday through Friday. How would that affect your job? Let the owner sign next to your warning. The owner also needs to agree to make all decisions in a timely manner. Think how you would handle this: you finished the tile on the job, your guys are waiting on the bathroom fixtures to get installed, and the client is still debating which style to buy. Time is of the essence and it costs money. Avoid that issue by outlining it in the contract. All design decisions and materials provided by the client must be ready when you need them.

Warranties should be noted as they may apply. Local laws may also require you to include them in a contract. Having a warranty in writing also shows how you stand behind your work. You should include a note stating that your warranty does not cover changes due to neglect and misuse, normal wear and tear, temperature changes and moisture, settlement, contraction and expansion, and other factors outside of your control. You may get perpetual callbacks to fix your work that keeps getting damaged by other factors.

Be specific on the conditions of the job site and expectations from your side and the owners' side. Generally you would want your site cleaned of all belongings before you start work. Write that down so the owner is reminded when they sign the contract. Your crew should never handle artwork, collectibles, and other valuables. Be very careful when you are asked to move furniture off and then back on the site. Take pictures of any existing damages before you begin. If the owner or designer is responsible for moving something out of the construction area, do not move it yourself, and avoid starting the work without it being moved. This is why it is important to mention it in the contract beforehand. You need to be compensated if the job gets delayed because the client or the designer fails to live up to agreed expectations. Your job site also needs to stay clean of owners' belongings while work is in progress. You cannot be responsible for the favorite blanket Mrs. Betty White left behind when she decided to read a book in the middle of your kitchen demolition after your crew left for the day. Guess what, Mrs. White? You went on the job and left a blanket that my guys thought was a drop cloth. When Mrs. White rereads the contract she signed, she will clearly see that she wasn't supposed to be on the construction site without you present and that she cannot leave any belongings on the work site. Owners need to be responsible for keeping pets and children, friends, and neighbors off the site at all times. Not only can they get hurt or worse, they can mess up your workflow by misplacing or misusing your tools. These are very important notes in a contract: remove all your stuff from my job site, keep your kids and pets away, I am not responsible for your stuff. (But maybe be a little nicer about it.)

Allowances need to be mentioned in a contract. There are certain things you can only estimate to your best ability without being certain of the price. If the project requires certain work dependent on other conditions, you should let the owners know they need to set a specified amount of money aside for it. Allowances are common and vary. Indicate them expressly.

You need the design part of a job specified in the contract if you are working with a designer. If you are working directly with the client, you need the design aspect written in the contract that much more. Clients expect you not only to build their specified work, but to also run around shopping for them. Oh wait, they also want several samples. Did I forget to mention that after you bought and installed that light fixture they do not like how it fits with the rest of the bathroom? *Mr. Contractor, can you please take it out, return it, go find me more samples, then buy and install yet another? All that work was a part of your bid, wasn't it? I didn't see it mentioned anywhere so I could only assume it is what contractors do.* No, it is not what we do, unless of course we are getting paid for it. Design work, meeting to go over samples, time spent on research, drawings and measuring, evaluating and planning, selecting and picking up fixtures, and writing up plans all take time. Specify this in the contract and mention the hourly rate you bill for such work.

You are not responsible for existing conditions. There may be a black hole hidden under the floor. There could be a typhoon lurking behind the walls. You do not know and do not want to speculate. There is absolutely no price given for an all-in job when you have unknowns. Don't do it, ever! Present your client with the fact that inside the walls you can find rotten pipes and wires improperly spliced together. You cannot predict what's to come, and you will evaluate the condition when it arises. Sometimes, the condo building requires you to replace risers when the walls are open, functioning or not. Such unforeseen variables are the sole responsibility of the client, as mentioned in the contract. Also, failure of existing work cannot be your liability. If you install a wall secured to a ceiling that is collapsing, repairing that ceiling should not be done on your time and money. Make sure after you start the work that any evident malfunction or existing problem is brought up with the client before you touch it, so you do not get blamed for it. If you ever think there may be obstacles in doing certain tasks, warn the client beforehand of possible scenarios. Always set expectations and present possible and probable circumstance. When you do find a faulty installation, present the solution to your client. Due to the cost associated with fixing old or bad work, many times customers elect to just leave it as is since apparently it was always fine hidden inside the wall. At this time, demand that they sign a waiver releasing you from the liability of not fixing the problem. You saw the issue, presented the possible danger, gave a solution, and they refused to fix it. All that needs to be written down and signed. Do not ever just bury a concerning matter in the walls while hoping for the best. Having the client deliberately sign a waiver will lay the responsibility on their shoulders, and oftentimes it will also get you more work when they realize it is rather serious.

Termination rights for both parties should be included in a contract. This may also be legally required in your area. If either the contractor or the client fails to live up to their expectation, the other party involved should have the right to end the contract and be released from further responsibility. Be careful, when you quit a job because you have gotten fed up with the client's unwillingness to work with you in a timely fashion, they will get another contractor to finish your job and possibly go after you for the difference. Be careful when writing this clause in your contract to protect your rights.

You obviously must include a price in your contract, but you need to also include the payment schedule. You need to check with your local consumer agencies about how much you are allowed to charge up front as a deposit. Charging too much may not only be illegal (thanks to all the scam artists out there), but may also raise some eyebrows from your clients. Charge fairly, but do not leave more than 10 percent for the final punch list. It is better you hold most of the money before you finish than your client having your funds as collateral; therefore, set that expectation in the contract. You need to establish trust, but you have bills to pay and need most of the money before tasks get completed. Be smart about getting paid, but also be mindful of the law. A 40%–30%–10% payment schedule is fair and gives you enough to work with. Most of your expenses are in materials up front and labor as you go, so you need most of the money early on.

The quality of work should be your priority, but you need to mention it in your contract to cover yourself from setting unrealistic expectations that you cannot deliver. Workmanship to be agreed on should be functional and aesthetically pleasing. You should relay that in your contract and include that your work will be of good quality. Promising *the best* or *superior standards* might get you into trouble when your standards don't quite align with your client's. When you, for example, read the instructions that come in a box of wood flooring, the manufacturer specifies that filler and stain are to be expected and acceptable practices during wood floor installation. You may get a client who will sit there with a magnifying glass and make you re-cut all your cuts, and call you back with the change of every season as the floors expand and contract leaving unnoticeable gaps. Also, be careful when asked to match existing finishes. Include in your contract the fact that you will try to match what is there to the best of your ability, but texture and color differences should be expected with the use of newer materials. Your client should understand that making the new look exactly like the old is impossible without redoing it all. Avoid endless demands by nagging clients who are never happy. When agreeing to a finish date, do not expect your payment at final completion. There may never be a final completion. Your client may have one punch list after another, keeping you on the hook forever. Your last payment should be at a substantial completion and only one punch list should be expected. You have the time to address what is on the list, and then you get paid. Obviously, you will go back and fix anything that was missed, but do not let your clients dictate that. Address in your contract the quality of workmanship, matching of existing finishes, the substantial completion condition, and the punch list.

I realize there are quite a few things to consider and include in a contract, but I can confidently assure you the above points are necessary. There may be even more things you need to think of on your own, but what I wrote here I have personally dealt with and would hate to see you go though the same unprepared. Though this excerpt is not about my personal story, I would like to share an account that may present you with the importance of linking all of the above in a contract:

"I was working on an electrical job through an interior designer I have done other work with and trusted. The designer agreed on the work with his client and I made my own agreement with the designer, where I never had to deal with the client directly. I had no signed contract, as I trusted my designer friend and saw no reason for concern; had there been a client misunderstanding, it would fall on the designer since he had the client relationship. Part of the agreement was to remove all the artwork and valuables from the room we were to work in. The designer agreed to move all of it himself. We would not touch any valuables, but only move large pieces of furniture. We agreed on a start date. I had scheduled my guys to start work, as well as my electrical sub. Five guys were expected to start work on this three-week long project: myself, one carpenter with one helper, one electrician and his helper. The week before we were supposed to work, the designer messages me with a change in plans. His client decided to stay in the apartment two more weeks before he left, so the work is postponed. Wait a minute! I now only have a few days to find my guys work, or pay them for lost time, and pay the electricians for lost time, as well as schedule everyone again for two weeks later, if they can fit me in their schedule in such short notice. This is a nightmare that's costing *me* time, messing up my future schedule, and most importantly losing me money! The client has no respect for what I am going through since his arrangement with the designer had no time frame references, and I had no written contract with the designer. So, of course, I will not be compensated for this loss. Ok, we move on. We finally start the work, after a few *more* delays from the disregarding client. The first day on the job, we realize that *nothing* was moved from the site. The designer did not do his part. With him there directing us, we spend about four hours moving most of the valuables out of the space. The rest of the pieces that are out of our reach, we just cover in blankets, paper, and plastic. We start doing the demolition work, but I am having scheduling problems with the electricians since they are now fully booked and cannot seem to fit me in their current schedule. We are losing time and falling behind schedule. The client is growing angry and threatens the designer. My own anger is equal at this point. To add to the frustration, while the client agreed to leave his home vacant for the three week scheduled renovation, he did leave a guest in his home.

This guy, who unexpectedly was now living in the space, would walk out of his room in the mornings, in his pajamas and slippers, take a stroll through our construction site while we were trying to use a jack hammer on concrete slab, and take pictures of what we were doing. He would nicely report all activity to his friend, the owner, who was away as he had promised. I'm infuriated, the owner is infuriated from 3,000 miles away, and the designer is just lost at this point. Again, there is no contract to fall back on; we are just taking it one day at a time with one surprise after another. To my horror, one day this guest points to a broken glass in a cabinet. He, of course, had already taken a picture of it and sent it to the owner, who is on the designer's case before he even knew what happened. It seems that one wine glass was broke. It wasn't even where we were working. I asked all my guys if they knew anything of it. They did not. No one ever heard glass breaking. I know my guys would run to me right away if they had broken something. We all stand there puzzled of how and when this glass got broken. As it turns out, this was a special collectible glass hand made in 1932, and brought in from Venice, Italy. The owner paid $600 for it a decade ago. The designer writes the owner a note apologizing for the broken glass, explaining how it was his responsibility to move everything before work commenced, but since the contractor was in the space, and probably broke the glass, that I would be paying for it. Lovely! So now I am paying for a broken glass, that no one knows who broke, and that was someone else's responsibility to move out of our way. Wait, how about the contract? Oh, right, I had no contract, so I had to foot this bill as well. For all we know, that guest who was nosing around after work hours, accidentally tripped in his slippers on one of our tools and broke that glass, while blaming us for it the next day. To make this story short, I am omitting other minor problems that I had on that job which could all have been avoided, or at least not expensed on my account, had I signed a contract with the designer. The lack of a contract turned that simple job into a labyrinth of costly errors. My single fault was to not have a contract, but all other mistakes fell on me because of it."

Change orders

This is a big item for contractors, and one that is scary for clients. Clients view change orders as sneaky ways for contractors to overcharge and make huge profits. Most contractors see change orders as ways to cover themselves from the unexpected and not have to foot a bill that was never accounted for when it comes. It is my experience that absolutely every single client will change their mind about something or will want to add new things to the project once they see their space start to change. I have also come to believe that almost always I will run into problems that were not on the original bid either because of issues with existing work or because the client left no allowance for unexpected circumstances. How will those changes affect your work flow, extend your timeline, and who will pay for them?

Change orders are a normal part of any construction project. They are to be expected and there is absolutely nothing negative about them. Let your client know this from the beginning, and mention it in the contract as well. Everyone needs to be flexible when it comes to construction. You have to be flexible with accommodating the needs and second thoughts of your client and your client has to be flexible with their budget to amend changes.

A change order needs to have its own contract that now references the original contract. It can be as simple as an email presenting the change to be made. Include the details of the change, the time this addition will take and how it will delay the project, and the additional cost. Some contractors charge an administrative fee for every change order; if that should be you, then mention this fee in the original contract. As with getting paid for your time, drafting and planning change orders requires time on your part, and you need to be clear on your expected compensation for such work. If you fail to provide why and what you are charging for beforehand, chances are your client might refuse to pay that bill.

When things turn sour

I hope they don't, but here are your last options for the worst-case scenario.

If things get really bad, stop the work. Show your client you mean business and cut your losses short. Before you do so, however, you need to try to mediate the situation and offer solutions to the problem at hand. If you cannot agree on things, or your client cannot decide on a solution to their problem, you need to stop any work. When you do this, you need to put it in writing. You need to clearly communicate that you have ceased all work until further guidance is given and payment in full up to that point. Write the solutions you have given, why they are appropriate, and why the client has refused them. At that point, either the client decides to compromise and get along with you, or they call another contractor to pick up where you left. If they hire someone else, they may just never pay you what they still owe and pay the other contractor in full, or they may try to sue you for the difference. Since you have it in the original contract (as you just read above), the option to terminate the contract is valid for both parties. Since they did not live up to their side of the bargain, the client was actually in breach of contract, so you may be safe from legal liability. You still may never see the money owed.

In case you have all the legal rights to collect your payment but the client refuses to pay, you have several options to resolve this dispute:

The first and most costly to you is to just walk away. The headache, time, and cost involved in fighting for that money might just not be worth it. You may just want them out of your life and the drama put behind you. Write them a thank you note, and erase them from your memory. As a good friend always reminds me when I am financially stressed, *I'll always make more*. Yes, you will always make more: more friends, more clients, more memories, more money, and more opportunities. Sometimes we just need to learn how to let go. There is always more happiness just around the corner.

If your wallet is arguing otherwise, you can put a lien on their property. Often called mechanic's lien or construction lien, such legal action limits the owner's ability to sell their home since it will show up on the title search and must be cleared before closing. Laws vary widely from state to state, but in general you must have added value to the real estate to place a lien on it. Liens also expire and may not ever pay you anything if the owners never sell their property.

The next step is to go to small claims court. Here you may present your case in front of a judge and get a quick resolution with minimal cost. The dollar amounts you can sue for vary. Kentucky and Rhode Island have the smallest limits at $2,500, where Tennessee has the highest limits at $25,000. If you are owed more than the limits, either settle for your state's limit or move on to the next steps.

Mediation, then arbitration. Both methods employ a third party to try to solve the issue at hand. Mediation is a non-binding discussion to end in a resolution, whereas arbitration is a binding process that judges your case. These dispute solutions may not get you what is owed but often strive to get to a middle ground.

Your last stand, and one which I hope you never face, is litigation. This process drains too many resources from all parties involved, except of course the lawyers. You should do everything in your power to never get to full blown litigation.

Legal options and limits are often included in contracts. Some states require them for validating disputes. Check with your local consumer protection agency as contract rules about the owner's rights to change their mind, lien, arbitration, and other legal matters may have specific necessities.

I usually like to believe that I am not one to look for trouble but that trouble finds me. My close friends and loved ones might disagree, arguing that I am a hazard to myself. I am generally quick acting. As quickly as I get tangled, however, I am able to untangle myself free and move on. Though you cannot foresee trouble, as you'd be a fortuneteller and have acquired millions playing the stock market, you can prepare yourself to battle trouble. Properly define your limits and expectations in a contract. Having this contract may just keep trouble away, just as those home security lawn signs keep the thieves away.

Chapter 8
Bidding

Risk comes from not knowing what you're doing.
—Warren Buffet

Most contractors are in this business because they love their craft. They are much more comfortable on the front lines than they are in front of a computer. They would rather apply their carpentry skills and compute numbers for framing a roof than be bothered with business math and other formulas. This is partly the reason why so many small contractors use a gut feel approach to estimating cost and bidding on jobs. Unfortunately, so many of them cannot stay afloat long enough using made-up numbers when charging for their services. You need to work at perfecting your business the same way you work at perfecting your expertise. Charging appropriately for your work is imperative to your well-being and company success. Too many contractors focus only on the work and neglect the business side. Bidding is where you need to get very precise and think like a CEO.

Many times, I have had the opportunity to see my competitor's bids on a proposed job. Either a client who preferred my company would show me his other quotes, or a new client who is trying to get the contractors into a bidding war. I am always curious how other professionals come up with their numbers. Some bids make me wonder how they can possibly stay in business offering such low prices, while other bids make me believe I am not charging what I'm worth. Why do bids differ so much between competing contractors? As we look at properly pricing jobs, we should keep in mind the process is an art and not a science. I am often reminded that some contractors are low-balling to just keep themselves or their crew working. During other signs of desperation, they would take any job that looked like it paid anything at all. Some contractors may use unqualified workers or illegal workers, keeping their labor cost much lower. Some may just be starting out on their own and are just downright clueless of what anything is worth or omitting to account for parts of the project. Some builders just use cheaper materials and inferior corner-cutting techniques; my heart goes out to those homeowners. I cannot compete with these contractors, nor do I ever want to. That would be a sure path to a frustrating career and eventual trip to bankruptcy court. At the other side of the spectrum are contractors pricing jobs that seemingly have an unlimited budget. They may have much larger overhead with extra support staff and added employee benefits. They could also be very busy working at that time, and inflate their prices because they don't really need the work but will take it for the right incentive. I have even had experiences when other contractors walk into a place and see dollar signs lining the walls (either because of the prestige of the building in which the owner lives, or because the client is a high profile public figure). Folks who are well off financially are no dumber or more eager to part with their money than your average middle-class American. Therefore, the really high and the really low bids will probably not be your competitors anyway. The bids that resemble yours are the ones you need to beat on cost, timing, quality, or all three.

Shooting from the hip and making up prices on the spot is like walking blind through Times Square. You are in the middle of the action, you hear the noise, you feel the excitement, but sooner or later, you'll get run over by a car. You don't want your business getting run over. You need to consistently make a profit and therefore properly identify your cost structure. Had you been the only contractor out there, and had the luxury of charging whatever you felt like, you could inflate all your prices and guarantee yourself success at the expense of the homeowners. The reality in a capitalist society is that competition makes for a better marketplace. It provides customers with better options and keeps service providers on their toes improving efficiencies. So while you may feel like a cowboy and throw your lasso out at freewill, you will get weeded out by a contractor who is pricing below you or by the other contractor who is pricing above you. Either your price is too high because you are fudging numbers to make sure you don't come up short, in which case you will lose most of your possible work, or your price is too low because you are the master of underbidding everyone, in which case you eventually run out of money since you cannot pay bills that were never accounted for.

There is only one exception to this rule of giving a price based on instinct. Many times customers want to know a price right on the spot, the first time you do your walk-thru before you even finish writing down all your notes. Giving a range based on experience can be beneficial for setting expectations. You should, of course, let them know that you need to sit down and do a proper evaluation before you can give them a clear picture of the project cost. People who have a dream marble bathroom and a $2,000 budget need to get cut off right away before they waste your precious time on writing a proposal. Being realistic with daydreamers can weed out ostensible cheapskates. They always want something "fast," "easy," and cheap. I thank them for their time, but that is not the type of work my business gets involved with. I've had an office-type lady ask me to show her how to skim-coat a plaster wall, as she'd be working by my side so we can cut cost. A reoccurring theme has also been for some economical clients to ask that we do all the work in secret without permits or the need to follow code. A contractor friend once told me a client offered to "feed his crew" so his estimate can fall within his budget. Go ahead and make up a price on the spot for these types of folks. Give them a price that will bring them back to reality. You will at least get some entertainment value out of it. I know right away that my price will be much more than the client is willing to spend, and frankly have no interest in negotiating down all my profit. If I want to work for free, there are plenty of projects I could be doing around my own home.

BUDGETING

You should always keep track of how your business is doing. You can learn something from the data. You can clearly see your financial picture though good or bad times. You can make predictions of future performance, and you can set goals.

A simple budgeting exercise would be to:

- Write down all your fixed costs
- Write down how much you want to make
- Add the two together—you must net that amount
- Now estimate how much revenue you need in order to make the above income

Let's run through the above exercise with realistic numbers. Let's assume you are just starting out, you are your only employee and you hire help as needed. The numbers will vary depending on where you live, but should still give you a good general idea. These are yearly figures.

- Your fixed costs: Liability insurance costs around $5,000 and worker's comp around $4,000. You

151

probably won't have an office and storage to pay for right away, and using your home as part of the business will yield some tax benefits. Your vehicle shouldn't cost more than $3,000 a year (if it is, you're wasting money before you can ever make any). Insurance companies report that the average driver drives 1,000 miles per month. Let's say half of that is for work purposes, so you'll be driving 6,000 miles a year. At the average pickup truck fuel economy of 16mpg, and gas prices at almost $4/g you'll be burning $1,500 worth of gas. Your average insurance and maintenance cost (again dividing it in half for work purposes) should be around $600 a year. Adding in portions of your phone, Internet, and other bills that partly get used for work, plus business accounting and banking fees: $900. Your total starting fixed costs should be around: $15,000.

- Your salary: I am constantly amazed at how little respect our industry gets from other professions. For some reason, the lady sitting in a cushy office thinks she deserves to make so much more than a contractor. She only works eight-hour days to our twelve-hour and weekend work. She most often doesn't have that much stress or responsibility for others. Her personal risk is minimal at work, while we risk everything we have on every job. Yet, she insists she should be making a six-figure income for her smarts, while she pays us $10 per hour. I simply refuse. I demand the same respect I give others for their work. I am worth at least, if not more, what most people make, while my responsibilities immensely exceed theirs. I deserve the same standard and quality of living of the average individual whose home I remodel. If I cannot achieve that, I need to close my business and get a job elsewhere. With that in mind, you should expect to be adequately compensated for your efforts. If you live near a major city, making $100,000 a year should be an achievable goal. This would be your salary plus profit; what you are left with after paying all your bills. At a 2,000 work-hour year, that's a pay rate of $50 an hour. If you live in New York City, you should strive for more; if you live in a rural area where home prices, cost of living, and ease and availability of renovations are much lower, you could be happy with less.

- From the above data, you must make $115,000 a year, whether you secure any jobs or not, to pay your bills as a business and make a living as an individual.
- You know your fixed costs. The work you take on throughout the year will have costs that are unknown to you. Materials, labor, and other variable costs need to be covered by your sales. If you are working in your trade, you will be saving one labor

had, and your net income should be upwards of 30 percent of revenue. With a $250,000 a year revenue, and a 40 percent return, you should be making your $100,000 pay. This, however, takes a lot of work on your part and a tight control of your costs. If you take more of a managerial role and only run your business without doing actual labor, your return might drop to 20 percent, and therefore you would need to sell $500,000 of work to keep your $100,000 pay. These are ambitious numbers. You need to set your expectations from the get-go.

Over time, you will be able to see your return rate by looking at past jobs with the data you have compiled. You will know exactly where you were profitable and what it takes to make the money you want. Always set budgets, set goals based on realistic numbers, and monitor performance.

Before you can budget and further offer bids to clients, you must first understand your *true* cost of doing business. You cannot just add up all the expenses on a job plus the cost of labor and use that as your estimate. Those contractors who neglect to see the rest of the cost cheat themselves and hinder their earnings. They constantly underbid and find themselves in a never-ending circle of struggling to pay bills. I am confident that with a little planning you will not be one of those contractors.

Your cost structure can be classified as follows:

- Direct Costs: These are the costs directly attributed to a job. All the materials, labor, and all other items you can point at that particular job.

- Indirect Costs: These are the items that cost you money just to stay in business and cannot be directly attributed to one specific job—your vehicle expenses, insurance costs, rent, costs to run your office, legal and accounting fees. Further, indirect costs can be fixed and variable. Variable costs could include the increase in marketing costs with recruitment of more work, the extra cost of insurance, or increase in maintenance costs from doing more business and needing more tools.

Overhead

Your overhead needs to be meticulously thought out. Every single cost you sustain that is not assigned to a job needs to be added. If you forget something, you will be paying for it out of your profit.

You need to assign a portion of your overhead to every job, in order to recuperate that cost. When you are first starting out, you may not have a track record and therefore find it difficult to assess your entire overhead and divide it into jobs. Work with the jobs you do have (if you have three months worth of past work, multiply those numbers by four to get your yearly figures) or estimate in the beginning. There are two ways you can recover your overhead and figure out how much should go into every job: on a percentage or dollar base.

Percentage overhead

To figure out your overhead markup, you need to know what percentage of your direct costs account for overhead. Add your total direct costs for as much data as you have, then add all of your indirect costs. Divide your indirect costs by your direct costs to arrive at the overhead percentage. Let's say for the past year you have had $300,000 worth of direct costs on your jobs. Your indirect costs totaled $40,000. $40k / $300k = 13.33%. You now know that for every single job you take, you must mark up the cost by 13.33% to recover your overhead.

Dollar overhead

Another simple method to account for overhead when charging for a job is to figure out how much overhead costs you during the time you are on that job. Take your total yearly overhead and divide it by 50. As above, $40,000 / 50 = $800. Therefore, for every week you are on a job, you need to add $800 to your price.

I have assumed 50 work weeks in a year. You may work much less during rough times. You may also work more if you have no interest in taking time off for vacations. Also, neither the percentage nor the dollar-based computations can account for the future with full accuracy. This is why it is important to keep good records and reevaluate your formulas. You may find you are not marking up your work enough and cutting into your profits. Also, it is worth noting that smaller jobs should recover more overhead relative to the larger jobs.

Labor burden

Your true cost of labor is not just the hourly rate you pay your guys. The labor burden makes up the extra cost a company incurs for having employees. For all employees and even some subcontractors, you need to pay additional worker's comp, disability, and liability insurance. For full time employees on payroll, you will be paying a share of social security and Medicare tax, unemployment insurance, health insurance, vacation and sick days, and possibly retirement plans, employee education, and supplies allowances. All of this adds up to 50 percent or more added cost to your labor.

If you just hired someone and agreed to pay them $30 an hour, realize that you are in actuality paying them $45 an hour. It is imperative to price the labor burden in your calculations.

An additional note on labor burden: general liability insurance and worker's compensation insurance are revenue driven. They will cost you more the more work you do and the more hands you hire. The insurer will audit you at the end of the year. They will ask for payroll records and other tax documents. If any of your subs did not have their own worker's comp, you will be paying for their share as well.

BIDDING

Bidding correctly on jobs is imperative to your survival. You can bid too low and have plenty of work, but go bankrupt since *you* are essentially paying homeowners to have work done. You can bid too high and still go out of business because you will get no work. You need to properly account for all variables and charge enough to pay your bills and make a profit.

Some of my guys are always asking me questions about the business side. They are good carpenters but admit to be clueless about how to manage a business or properly estimate cost. They tell me horror stories about side jobs they took where they gave a client a price, then halfway through the job they ran out of money and had to quit. They sometimes come up to me with some really smart pricing tricks they heard from other contractors—seemingly clever and easy formulas that take the price of the material to be installed and multiply it by four, which would give you a fast and accurate price for the job. I just smile and explain the many reasons why that method will get you into trouble and think in the back of my head, *no wonder you work for me and not for yourself.*

Profit should be a part of your business plan. Profit is good. We live in a capitalist society. No business would ever operate if it weren't to make a profit. Would you ever foolishly expect to buy goods at the cost it took to make them? I don't think so. Should Starbucks sell us coffee at the price they paid for the beans in South America?

If they did, they would run every single competitor out of business, and then they would go bankrupt themselves. There would be no more coffee vendors. These strategies just wouldn't work in society. This is part of the reason communism failed economically. With no profits anywhere, there is no drive and no progress.

You absolutely need a profit. You need to be compensated for taking risks and for having such elevated levels of responsibility. Your salary pays for you to work in your business, while the profit compensates you for the headaches of running the business. You should never drop your price below the point of making a profit. Figure out on your own what you need to make for a profit. Many contractors mark up their prices by 10 percent to account for profit. Most contractors, however, see their profits diminish to less than 5 percent of project values, due to unaccounted job aspects. The erosion of profits makes the need for them in your bid that much more important.

Putting together a bid involves two steps:

1. Estimate direct costs. These need to be concrete, hard numbers based on accurate quotes of material prices and labor times.
2. Make judgment calls about overhead and profit. Use either percentage or dollar markup for overhead, then mark up your profit.

Some contractors use a "per square foot" formula to calculate the cost of certain jobs. Those methods may be of use in new construction where all the materials and techniques are standard, yielding the same results. In remodeling, however, pricing tasks on a "per square foot" basis is fairly worthless. Two jobs of the same floor surface may be using a different design or layout pattern, could use different materials, and, how fast can *your* guys complete that task? To get an accurate estimate, you need to analyze all the costs and stay away from standard formulas. You may ask: "How about sanding a wood floor or installing baseboards? Those are standard tasks." Take a small square room: it has four inside corners. What if there are structural beams in each corner that have been drywalled over? You now have the same size room with eight inside corners and four outside corners. Do you think that would make a difference in the amount of work and material you need? You best bet it will.

When sending a bid for a job, you need to prepare a professional proposal for your client. You should ideally get back with your quote within two to three days. For larger jobs when involving subcontractors and fitting many different parts together, you may need a week to two weeks of work to get everyone's bids and work on your own. You need to stay on top of your subs when bidding. They are busy and most are also sloppy business people. In order for you to look presentable in a timely fashion, you need to remind them how time is of the essence, often. Words of advice on getting quotes from subs: don't abuse them. As often as possible, try to involve them in the process as little as you can. Calling a contractor time and time again and asking for them to spend the time on giving you quotes, and then never giving them any work, can make them sour and unwilling to cooperate. If your electrician charges $100 per switch installed, do not call him every time you need a few switches and outlets put in, have him go to the site and put together a bid. You already know your cost, use it and call him only when you have the job. When you do need to get your subs on the site to see the proposed work, make sure you make copies of the plans and let them keep their own sets. The benefit of having subs walk through the site with you is that they will often see things in their expertise you might miss.

The proposal should always be written on your letterhead. You should have a template saved to use over and over. As discussed previously in the branding section, all of your written proposals should have your company logo, contact information, licenses, accreditations, and affiliations. You need to date it and make it personal by titling whom the proposal is for and signing it at the bottom with your name. The proposal needs to be then broken into several sections that follow the project in an easy-to-understand and organized fashion. You may separate sections by categories, such as electrical, plumbing, painting, demolition, or by areas to be renovated, such as kitchen, bathroom, lobby, or just in any other way which makes planning easy for you. Remember you cannot overlook any detail of the job. As you go down the list you made with your client, write in your proposal every line item and give it a price. Sending one price for the entire job without showing how you came up with it is not good professional practice in this industry and will leave your clients wondering if you know what you're doing. Remember to include a section detailing what is included in your price and what is expected to be purchased by the client.

General conditions

The first or last section of your proposal should be left for general conditions. These are all the items that cost money and support the work to be done on the job. These conditions make the work possible, but they are not items the client asked for. It needs to be evident in the proposal that you cannot work without these conditions.

- Insurance and permits: Going to the department of buildings, filing and waiting for a permit takes time. Your time costs money. Filing requirements by your insurance brokers also takes up your time. There are times when the job calls for added insurance, which directly costs you money to upgrade. All these costs and time commitments translate to line items in your bid.
- Pickups, deliveries, and storage: The mobilization of tools, pickup of fixtures, deliveries of materials, and storage of items on and off site. Even parking costs where applicable need to be accounted for.

- Site work and prep. Preparing the site for construction takes resources. Moving furniture, sealing off areas from the work site, setting up protection for walls and floors, as well as cleaning up at the end of every day and setting up in the morning.

- Planning fee. As you wouldn't expect anyone to work for you for free, it shouldn't be expected of you. Your time to plan, draw, design, shop for fixtures, present samples, and other background work comes at a cost. Charge per hour, per item, or as a percentage of work. Clients who cannot afford a designer but expect you to do all the design work, for free on your own time, need to be brought back to reality.

- Management. When applicable, hired management for that job site should be priced as a line item under general conditions.

- Error margin. This is not an area where you can freely pad all of your sloppy estimating work. This should be an allowance set for unexpected costs. Problems beyond your responsibility will arise and the client should be prepared for additional charges. Instead of adding extra fluff in every task to make up for this margin, you can be honest with your client and explain how you are keeping the line items at the real cost with the expectation that this allowance be available. Remember, there are also added costs you will incur from future items such as callbacks, breakage, touch-ups, systems you must install that need to be learned by your crew, and sloppy design that you must rework. Envision what *could* happen and plan for it financially.

- Overhead and profit. Of course, your client needs to realize the price associated with keeping yourself true to your real costs (overhead) and staying honest about why you are in business (profit).

Now that you have all the pieces together in your proposal and you did not overlook *anything*, you can go ahead and add up all the numbers to get your bid. That is your client's quote and you should stick to it. You worked hard enough on putting it together, so don't throw it out the window at the first sign of negotiation. Instead of seeing how you can lower your price, try to see how you can lower the workload. Clients always seem to think we don't quite need all that money for the renovation, so we should just assume they don't quite need all that work to be done either. As fast as clients expect us to let go of some of our profits, they should be ready to let go of some of their ambitions and make the project smaller to fit their budget. It is a hard reality for most to comprehend, but it makes perfect logical sense if they put themselves in our shoes. I wonder how most people would feel if their employer started negotiating down their salary.

Nevertheless, clients will try to argue what your work is worth instead of debating what they can't afford. Never accept a project under your cost. Ever! You will absolutely regret it later; numbers do not lie. You know your expenses clearly. You meticulously went over all your material and labor costs, as well as extra items and overhead. That is your bottom line, your break-even point. You shall never commit to a job below that price. There is no: *what if I can pull it off for less?* Ask yourself: *what if you can't?* Walk away!

The only way to negotiate your prices down and agree to take on the project at a lower price than quoted is to reduce the factors that are estimated and not hard numbers. I would highly recommend against doing so, but hard times may find you in need of making hard decisions. You may try to trim your overhead, but be very careful not to eliminate it from your calculations, otherwise you will have no money to run your business. You may reduce your profit or even take a pay cut for your own salary. There is a point though when your incentive is so low that it cannot be worth taking on all that risk. The last thing you can do is to reduce profits for your subs. Communicate with them the need to lower prices to secure this job. You will reduce your markup and they should lower their profit to help you close on this deal.

The last thing I will mention about quotes is that clients often want them on the spot. They want to know a realistic price range at your first meeting. As mentioned earlier, doing so may weed out most bad apples. Having a workable range also helps the homeowner be prepared for the actual bid. From your data of past performance, you should have a general idea of what a project of that scope should cost on a weekly basis. If most of your projects that size and format cost about $5,000 a week, it is probably safe to say this is your range. So just do a quick estimate of how long this project should take and give your client a quick and rough range. A two-bedroom apartment full gut renovation will never take less than two months to complete or cost below $50,000. That is realistic. If the client only has an approved loan of $40,000, then it is time to go back to the drawing board and trim down the scope of work.

COST CONTROL

When asked how much money they actually made last year, most contractors will give you a number off their tax return. They honestly have no clue of the exact figure, even though they may be disillusioned into thinking they know. The accounting gets complicated and all the money spent gets blurry. Did they really save all their receipts? How about those coffees you got for the guys that one time? How about that $20 tip you gave to the delivery guys? Did you account for the parking tokens or tools? How about the cleaning supplies you took from home to help clean up that mess at the job site? All the little overlooked things add up to a big difference over a year period, and you won't have a clear understanding of your profit. You probably don't get to keep as much money as you thought. Even if you are living comfortably, you might not be getting suitably rewarded for your efforts. You have no way of knowing unless you are ultra strict about keeping track of *everything*.

When it comes time to making a larger profit without charging more, there is *always* room for improvement. I am certain you can always keep a tighter control and slash your expenses. Are your guys using all the paint down to the last drops in the can? What if you have a few cans with a little paint left of different colors and finish that are now a waste? Why not mix them all together into primer? Don't throw anything out that might be usable. Are your guys making too many cuts and throwing out material? Are they mixing too much plaster or thin-set and it dries before they use it? You might be surprised at the waste you find in the garbage if you start going through it. Are screws falling out of tool belts, then swept up with the dust? Are the glue tubes being left open and drying out overnight? Are countless rolls of tape tossed around half used? Is the grinder always left on the floor so that someone else steps on the blade and breaks it? Being careless with your materials is not acceptable. You need to set a standard for accountability and responsibility when using your materials. Also, ordering too much material and having it all sit on the job site will give your crew the impression they can be as wasteful as they wish. Why use those last two small pieces of sheetrock when there is always a new sheet to cut from?

You need to keep track of your ordering and use of materials. Know what you ordered and track how the material was used. Did you end up needing to order more?

Why? Was your calculation off or was the material used improperly? Figure out those metrics and document them for the next project. Your bids are based on expected cost of materials and labor. If you fall short on either, you are losing money. Make sure you are delivering what you had expected. If your crew is not, you need to adjust your estimates and/or educate your workers on what is expected.

Aside from keeping a very close eye on your materials, you need to also monitor the labor. Are tasks being completed in the expected time frame? Why not? How can you be more efficient? How can your crew work better and faster? You sometimes need to work alongside your crew to train them, establish standards, and demonstrate how to improve.

Be the Jack Welch of your own GE. Be brutal at cutting inefficiencies. Always be honest with yourself, honest about your performance and that of others, and honest to those around you. This is your business, and you are in control of its success. You are also ultimately responsible for all that your company puts out there and the consequences it creates.

if you made money on that particular job. All your numbers should follow what you have included in your estimate and bid. If they do not, then you need to see how you must improve your estimating. You have essentially calculated your gross profit on every job (everything you made on the job, minus everything you spent on it).

2. Account for all overhead in a separate spreadsheet. All of the costs associated with running your business and bringing in work need to be tracked. This sheet is what will get you your overhead that gets marked up to every job.

3. Have one master spreadsheet that includes the summary of revenue and direct cost of every job and the overhead. This will be your financial performance at a glance.

Summarizing the above 3 steps:

Revenue

<u>- Direct Costs</u>

= Gross Profit

<u>- Overhead</u>

= Net Profit

In the direct costs sections you would generally want to include your salary for doing work on the jobs. Under overhead, you would want to include your pay for working for the business. Let's stick to the earlier example and assume your hourly rate is $50. Log your hours doing paperwork—that goes under overhead. Log your hours doing labor on jobs—that should be a direct cost. Now you are getting a very clear picture of your *real* cost and *real* profit on jobs and overall in your business. From these numbers you can come up with ratios. What percentage of revenue is your gross profit? Are you seeing discrepancies? Why and so forth . . . If you marked up your profit on bids by 10 percent, you will most likely start seeing that you can't keep all 10 percent by the end of the year. Most contractors have a net profit margin of less than 5 percent. Some of it gets deteriorated by callbacks or other mistakes you made. How much? How can you do better? By keeping solid records you are able to track all such statistics, spot problem areas, and come up with solutions.

Banks will ask for your tax returns as well as financial statements. The two statements widely used are the Income Statement (income − expense = profit), and the Balance Sheet (assets − liabilities = equity). It is a good idea to complete these financial statements at year's end. Corporations have to file them for tax purposes in one form or another. Financial statements carry more weight when they are audited by a professional. Bonding agencies will also require them, as well as any government contracts you may decide to bid on.

I wish you never have to go through an IRS audit; however, you must always be prepared and armed with the best strategy. Always keep all receipts and all supporting documents. Remember, just because you have a receipt for an item, it does not prove that you actually paid for it. If the credit card statement matches your receipt, then you have a proper track record. If you have a bill from a vendor but no proof of a cashed check by them, then you never paid it. You paid it in cash? Great, let's see that cash withdraw slip. The cash doesn't seem to have come out of your accounts, so it looks as if you have other income you are not declaring. Cover all your bases and keep proof of all transactions.

One other topic to stay clear of is the personal use of business accounts. Always keep your personal and business accounts separated. Not only will it make your life easier when reconciling accounts for your bookkeeping, but it will keep everything legally intact. If you are making personal purchases with your company credit card and writing business checks out of your personal account, your limited liability is out the window. Your business is one and the same with you as a person. The government also doesn't seem to be so lenient when you seemingly personally benefit from the use of pre-tax business assets. It is really easy to get in the habit of using the separate accounts for their intended purposes.

FILING REQUIREMENTS

Tax forms are confusing. The language is puzzling. Even the instructions on how to fill out the forms are arcane. It's enough to make any business owner want to quit. There is, of course, the option of hiring professionals to take care of all your accounting needs. That does cost money, and when you are starting out you'll need money the most.

There are quite a few taxes you must file, and they must be on time.

To the federal government, you must file payroll taxes and send withholdings on federal tax, and employer's portion of social security and Medicare, on a monthly basis. You must also file corporate taxes before March 15, and personal income taxes before April 15.

To the state and local government, you must report payroll taxes and send in withheld money for state and city income tax, unemployment insurance, and state disability or reemployment fund tax, on a monthly basis. If your state imposes a sales tax, you must also file that tax and send that money in, sometimes not on a calendar year basis.

The good news is that all of these taxes can be filed online and the money automatically deducted from your account. Registering on EFTPS for the federal portion and on your state's finance website is simple. Also, online you get to answer clear questions about the numbers you are inputting rather than filling out strange unclear forms. Everything is easier online with step-by-step wizard templates. I would recommend shopping around for a professional service if you find taxes get too complicated and time consuming. My bank, for example, does all my payroll filings and deposits at no charge, so that option does exist.

Chapter 10
Benefits Planning

If you live each day as if it was your last,
someday you'll most certainly be right.
—**Steve Jobs**

Your business is set up artfully. You have stellar workers. You are booking jobs and completing them on target. You timely handle all the office paperwork. You are living the self-employed dream. But are you ignoring yourself in the process of taking care of your business? When your business is thriving, you need to plan your own personal future. After all, if you do not take care of yourself, who will take care of the business?

For a prosperous business, you need to look at your current needs of booking jobs and making money, but you must also look further down the road to medical provisions, savings, and retirement. Having a solid structure to your business is important for your success. Having something to fall back on during turbulent times is essential for your own well-being.

HEALTHCARE

It is very unfortunate that the greatest country on Earth cannot take care of its people. All the debating in congress has still left many Americans hurting for healthcare. I do not have a solution to this problem, but it saddens me to see that even third-world countries can offer their citizens free health benefits while we argue and watch our society suffer. I am very patriotic and love my country for all the opportunities present here, but I am not a blind nationalist. Our issues are not getting addressed while insurance companies and medical facilities reap the benefits of a system that has failed many of us. Many young people are uninsured and have no access to medical help should they need it. Many people unfortunate enough to have serious illnesses are bankrupt from their medical bills. *You*, the entrepreneur, are trying to figure out how to insure yourself while still making a living. This is a problem.

Due to the nature of their work, contractors are susceptible to sudden injuries and long-term health complications. It is crucial you have a backup plan should a health concern arise.

Your first step in staying healthy and hopefully never needing medical attention is to live a healthy lifestyle. I am a firm believer in having healthy habits, not dieting, no quick fixes, not changing who you are, but just making smart choices. Eating healthy and exercising should be a part of your life. Take care of your body and it will happily carry your mind into long healthy years. Regardless of how busy you are and how hectic your life may get, you can always incorporate your healthy living into it. You can eat anything you want, just do it in moderation. If you have no time to exercise, incorporate everyday fitness steps into your daily routine. Take the stairs rather than the elevator, ride your bicycle to work when you don't need your truck, install a pull-up bar in a doorway and do a few pull-ups anytime you walk under that door. All of these and many more creative ways you can stay fit do not have to take extra time out of your day, but just be a part of your every day. If you have never eaten healthy or exercised, and you plan on starting an entire regimen, chances are you will fail. You cannot disrupt your life and expect it to change. You need to incorporate small steps that can become routine. While this is not a fitness book, remember to keep yourself healthy before you ever have to use this dreaded health insurance.

Regretfully, even the healthiest person cannot live without insurance. You just never know what the future brings. While you are working hard to make money and secure yourself a comfortable lifestyle, unexpected hospital bills can simply bankrupt you. In extreme cases, people have not only lost their business but also their life savings and homes over health obstacles. This is not meant to frighten you but only to motivate you into being prepared for anything. I remember when I first joined the Army, a drill sergeant said, "The reason we have the best trained and equipped army in the world is so we never have to use it." Over a decade later, having been disappointed by politics and having gone to fight a war overseas, I take that theory with a grain of salt. I do, however, hope that your best health insurance plan never gets used.

In recent years our nation has made some progress in offering every citizen the opportunity to be insured. From the highly debated and intricate Obamacare to the state mandated insurance rules, we have more options for coverage. Still, health insurance is costly and superior coverage is difficult to obtain.

A few places to try getting insured are:

- Look for medical insurance brokers. Usually these agents are well informed and have many connections to insurance companies. They can help you navigate plans and offer options for coverage. They can also shop around to find better pricing or set you up in a group.
- Look at labor and trade associations you can join that offer incentives such as health insurance. The benefits they offer may far outweigh the membership dues. Group insurance is much cheaper with better coverage than individual insurance. The insurer can spread their risk over a group of people to their advantage and your benefit.
- Ask other contractors or independent business owners about their health scenarios. You may even be able to form a small group with several of them.

- Check with your state or the federal government for options available to self-employed individuals. New laws are making insurance more affordable and accessible to everyone. Also, be inquisitive about laws requiring you to provide insurance to employees. Some states are mandating health coverage to your staff.
- Check with your spouse's employer. Oftentimes, there may be options to add the entire family on an existing insurance. (If you have no spouse, maybe it is time to tie the knot. I suppose love should have something to do with it as well, but health is more important in this chapter.)

When looking at insurance policies, research and understand your options. Do not assume you are covered under all circumstances and for all procedures related to your health. Compare plans based on cost versus quality of service. The cheapest plan may not get you much when you actually need it. Analyze deductibles, co-pays, maximum limits on reimbursements and liability. Look at what is actually covered and what is not. Read reviews online on specific insurance plans, on the doctors you are considering, and reviews on networks and hospitals. Be as informed as you possibly can so you do not take a chance with your health.

Do not forget to look into dental coverage. Depending on your dental history you may need it more than others.

I, for example, due to numerous accidents and damage caused by my ignorance or unfortunate events, would not live without dental coverage. Remember that your smile may say enough about you at a first client meeting. Add to a positive first impression and take care of your pearly whites. For personal pride, I am not looking forward to dentures in my later age, so I am doing my best to keep whatever is left of my own teeth going strong.

Medical costs are generally tax deductible. Your tax benefit may be greater depending on your business status and your medical-related expenses. This is an area where the law is tricky and can be interpreted creatively by accountants. You should definitely have a conversation with a tax specialist about your insurance and other medical expenses.

Other areas for you to review are disability and life insurance. Both of these topics are really sinister, and, while you do not prepare for them to happen, you should plan ahead for that random risk.

In construction, the possibility of becoming disabled is very real. Disability can mean your inability to work partially or fully for a few weeks or a few decades. Review disability insurance with care and try to plan for the unthinkable.

Life insurance may not make much sense if you are a single person with no immediate family of concern. For the rest of you, in the unlikely event that you will not be able to provide the lifestyle you created for your family, you should consider what is left for them if you are gone.

SAVING AND INVESTING

I cannot stress enough the importance of proper personal financial planning. You cannot successfully run a business if at home your finances are in shambles. You must always monitor your business and personal long-term and current bills, debts, spending, and savings habits.

One of the reasons you are going into business for yourself is to gain financial freedom. You may have your own definition of what it means to you, but being debt-free and not worrying about the next paycheck should be at the top of the list. Maybe you currently have college debt, maybe you mounted more debt to finance your new business, and maybe your home equity loan was used for new tools. Regardless of how much and what kind of debt you have, it should all go away at one point. Make it your goal. You do not have to call Suze Orman to understand why paying down your debt is important. When you have money earning you 0.5 percent interest in a savings account but your car loan is 10 percent, you need to stop paying the banks and start paying yourself. My grandpa used to say "Don't pay the bankers, they make enough money on their own." While World War II was certainly a different time for bankers, today I still agree that paying interest is silly if you can avoid it. Of course you need to have some savings for running your business, but beyond that, you should be paying down your debt. Debt keeps you enslaved. Being debt-free means you have no obligations to anyone other than yourself. There is some debt that actually advances your life goals. Owning a home is rarely done without some sort of financing, but even there, it should be fully paid off in thirty years or before retirement. Pay unnecessary, higher yielding debt first, then budget the rest of your debt until it is all gone. Waking up every morning knowing you owe nothing to anyone is a liberating feeling. Work toward that, and someday, all the money you make will go toward bettering your family and your charitable causes, and not to advancing the banks.

The national savings rate is currently hovering around 5 percent. Since 1959, on average, Americans have been saving around 7 percent of their disposable income. You can do much better for yourself. Once you periodically start putting money aside, it becomes habit. Save as much as you can for rainy days and for paying yourself later. A key to having a sustainable financial lifestyle is to live under your means. Do not fall in the trap of living above your means and spending money based on hopes of future income. Your future income is never certain, so spend and save today as being cautious for tomorrow. Buying personal wants on a credit card and paying it back later is a devastating path to failure. Prioritize business and personal needs, and place savings before wants. You should strive for a positive equity slope. As you go through life, your worth should constantly increase. You shouldn't try to keep up with the national average or any other statistic, but to maximize your own savings plan. Setting and living up to this goal will provide you with self-pride knowing that you are on a successful savings path. At the end of each year, you should see an increase, as small as it may be, in your home equity, personal savings, retirement savings, investments, and other assets. This is not an increase in the market value, but in the amount you contribute every year. Make savings a priority and curtail other activities when you feel there is nothing left to put aside. Make your coffee at home in the morning instead of getting it from Dunkin', or pack your own lunch instead of going to Subway. Take that extra dollar you saved and throw it in the piggy bank. You can always find a way to save. Act as if you care about your future now, and you shall live a happy one.

Once you start putting some money aside, you need to consider investing it.

We could all learn a little lesson from old-school gypsies. Due to lifestyle and uncertain times, any money they would earn from trading and manufacturing goods for sale, gypsies would use it to buy gold coins. Those gold coins would not get spent, but saved for the future. They generally lived fugal lives and kept buying gold coins with their savings. The gold coins appreciated with time and held their value with inflation. Buying gold was probably the only investing gypsies ever knew in those times, but those coins secured them a future and many were passed down to the next generation. By buying those coins, they were able to put money aside fast without the fear of spending what they had. Today, you have many other investing choices, but the moral remains the same: periodically put money aside into a safe vehicle.

While I had to take the Series 7 and Series 63 exams, investing advice is not the purpose of this book. Those exams administered by the Financial Industry Regulatory Authority are required for individuals selling financial products, and I have long left that business. I can, however, tell you to be very careful when investing. You should not get involved with things you do not fully understand. It is far more important for you to keep the money you have worked so hard for than to risk it in hopes of making more. It is too easy for someone to open an online broker-age account, transfer their savings, and then gamble it all on some alluring stock they saw on TV. The promise of overnight riches is hard to ignore, but I know you under-stand perseverance and the value of money. You are a hard working contractor. Getting involved with fancy financial products may leave you disappointed. Start investing early on in your career. Start with basic and safe products like Certificate of Deposits and Government Bonds. While these vehicles offer low returns, you will sleep at night knowing that your money will always be there, growing at a slow pace. Corporate bonds or stock of major companies that are most likely not going bankrupt in the next dec-ade are another good option with slightly higher returns. Getting involved with up-and-coming stocks and high-yield bonds should be kept to a minimum. Only risk what you are willing to lose. You may have bought a pharmaceu-tical stock that promised to cure cancer. You wake up the next morning and half of your money is gone, as the drug did not pass FDA approval. Would you be all right with that? Probably not, and since this is a realistic scenario, I would take stock market investing seriously and avoid the unknown. Trading leveraged and derivative products is absolutely not advised for the everyday retail investor. While I am not a big fan of financial advisers, they do play a good role for the inexperienced investor. Get recommendations and ask for their track record before hiring one to manage your money. Also ask exactly how they get paid. If they make a percentage of your assets, they have no incentive to monitor your money closely; they make money even if you lose. If they get paid on performance, they will have your best interest in mind. For the DIYers, there are mutual funds that allocate your assets based on your target retirement year. Investing in those will offer diversification and piece of mind, while essentially doing all the things an adviser would do, except for a lower fee.

Other forms of investing include tangible assets. You may look at buying property for future rental income. You may want to invest in a joint venture and open a local laundromat. Maybe you will consider buying artwork or other collectibles. Again, use your judgment and get involved in an investment opportunity only when you fully understand it. Generally, hard assets will hold their value against inflation over time, and may provide you income in the future while locking your savings away from your spending hands.

When saving, reflect on major life goals and plan accordingly. If you have children, think about their college education and consider tax deferred college savings plans. If you plan on buying land in the woods and building your dream vacation cabin, set up a separate savings account for that endeavor. Maybe you love to travel and have firm plans to go around the world, or if you are an extreme adventure seeker like me, maybe you want to climb the seven summits. Such goals are not only demanding and time consuming but also very costly. You need lifelong planning and saving in order to see your dreams come to reality.

RETIREMENT PLANNING

Retirement is a giant step in life. Some view retirement completely different than others. Some want to work as hard as they can, save up as much as possible, then stop working as early in life as they can. Others view retirement as the average sixty-five-year-old who leaves a job and starts collecting social security. Some people retire from their long-term job, but work through retirement in one form or another. If you are like me and cannot sit still for more than ten minutes, you will probably never be able to completely stop working. The key is to make sure when you are ready for retirement that you don't just retire *from* something, but retire *to* something. Pursue your hobbies and have your nest egg ready to support them.

You should think about your retirement way ahead. Start saving *now*. It is absolutely never too early to start saving for retirement. I wish I started in my teens when I got my first job, had I been smarter than I thought I was. Your dreams do not just happen on their own. You are in charge of turning them into reality. Think about that dream home you always wanted to build for yourself. How about the journeys you'll be taking with your grandkids? Remember all those places you have been wanting to see but never had the time? How will you make all that happen?

When you retire, you cannot rely on savings and investments alone. Many people make the mistake of just putting money aside and assuming they will use it in retirement. The problems with that mentality are: the money you are saving is taxed at your current rate, the interest and capital gains you make on your investments are also currently taxed, your savings are easily spent in case of an emergency, and you may take higher risks with money that is not locked away for the long term. Contributing to qualified retirement plans gives you tax benefits and you may think twice before taking money out since there are costly penalties.

There are a few options for retirement accounts: Individual Retirement Arrangements (IRA—ROTH or Traditional), Simplified Employee Pension (SEP), Savings Incentive Match Plan for Employees (SIMPLE IRA), and solo-401(k). You should research the specifics and decide what is best for you based on your tax situation and company structure. You should, however, open an IRA right away and start contributing the maximum amount *today*. Additionally, you may be able to open other employer funded plans as mentioned above, considering income and contribution restrictions.

I'll go back to our friend Warren Buffet since his basic American, down-to-earth, common sense smarts are appropriate for this section. *Someone's sitting in the shade today because someone planted a tree a long time ago.* You are responsible for the shade you should be sitting under in retirement. That tree needs to be planted by you, then watered frequently, fertilized, pruned, and cleared of weeds. You need to make sure your tree grows with you.

Depending on your chosen lifestyle, your fixed income may or may not support your living into later years. Social security as we know it will certainly not be enough for you to live on; if we even get to have social security by the time some of us retire. Your supplemental income will come from your individual retirement accounts, your savings, and your investments. Even so, you should consider retiring partially by working part time. You can use the skills you have acquired to work as a consultant, an estimator for another company, or to work at a local hardware store. You may even look into teaching. All of these activities will give you a sense of pride while keeping you busy and providing extra income. You can generally choose your hours and work as little or as much as you can. Be mindful that your age and income level will impact the amount of benefits you get from the government. It may be too early to talk about exact social security and Medicare benefits, but you should set realistic expectations and know what you get when you approach retirement. If by the time you retire you have a younger partner who wants to take over the business, you may have leverage to strike an agreement for you to get paid a yearly fee for leaving him the business. You may also consider selling your assets and the long list of happy clients to a larger competitor. Those are also great options for extra income coming from your business when you are ready to retire it.

You must pay off all your loans before you retire. You need to have no such burden or extra expenses. Your income in retirement needs to pay for your future living and not for your past. You need to be able to enjoy the fruits of your lifelong labor. This is the time when you should be selfish and put yourself first. There should be no reason for you to compromise your retirement in order to leave your heirs more. Live for yourself! I realize you love them and want them to be better off than you ever were, but give them a chance. Let them achieve their own dreams and live their own realities. You made it on your own, and so will they. Enjoy your retirement to the fullest; you worked hard enough for it.

Though it may be early to plan your actual daily retirement life, remember to start dreaming and planning your retirement as early as possible. When you get closer to your chosen retirement age, you should visualize what you would be doing with your time. Some retirees fall into depression and find no useful value in their lives. By visualizing your retirement, planning it, and setting goals, you will live a happy and meaningful life after your working days are over. Take time to teach others what you have learned. Pass on your knowledge to the next generation. Enjoy nature. Take long walks and feel the ocean breeze at sunset or the cool mountain air at sunrise. Enjoy your place on this vast beautiful planet we share. Resolve any misunderstandings you may have or have had with others in the past. Forgive and ask to be forgiven. There should be no reason for regrets. Wipe all anger and resentment by addressing it. Being happy is a choice we have in life. Make your choice daily.

Chapter 11
Good Luck!

The question isn't who is going to let me;
it's who is going to stop me.
—Ayn Rand

As the book comes to an end, it marks the beginning of your triumph. It is now *your* time to go out there and make *your* journey count. The average American spends thirty-four hours a week watching TV. How will you maximize your time? I hope you remember most of the advice found in these pages, and when confronted with those scenarios, you do not make the same mistakes I made. Sometimes the best teacher, however, is a mistake. So if you do happen to stumble, there is no need to dwell on what tripped you, but only look ahead at where you are headed.

Business success takes meticulous planning. I am confident you can accomplish your objectives by looking ahead while having expectations and setting goals. Where do you want to be and how will you get there? Figure out what you

excel at and where your weaknesses lie. Hire the right people to complement your skills and allow them to be proud
and responsible. Remember that you are hardly ever alone
and you never have to reinvent the wheel.

You need to put yourself and your positive message out
not only when you need the work. Reaching out to people
should not be done out of desperation but in hopes of building relationships. Do some free or charitable work to prove
yourself. Regardless of what you do for work, you should
always be proud. An honest living is a great living.

As you shall find out, your most successful sales leads
come from selling you have already done when you weren't
even thinking about it. Those who know you previously
will trust and support you if you haven't let them down.
Your best customers will come from circles you or your
friends are involved in.

Join a club. Demonstrate you can be a leader.
Stay involved in your local community.

Remember your clients are your lifeline. The reason
you get to enjoy your lifestyle is not because your services
are so spectacular they cannot be found anywhere else, but
because your customers chose to use you. The inspirational
common sense investor Warren Buffet wrote in a letter to
his shareholders:

> Decide early in life to make your money by selling things
> that you really believe are good for the customers. Make
> this a rule before you write another word to your read-
> ers or offer another product to your customers. Life is

too short and your reputation too fragile to not have your audience first and foremost in your mind and in your heart. Rules like this make it very difficult to lose.

Butterfly Effect

In mathematics, there is a theory that presents how sensitive and dependent various systems can be to initial conditions. The Butterfly Effect analyzes how very small changes in an initial state can have large impacts on future states. A tornado, for example, starts out weeks earlier as a very small change in weather conditions that can be equivalent to the wing flap of a butterfly. A ball left at the top of a hill can roll into any one of many valleys depending on a very slight difference in its initial position. Your choice to hit the snooze button this morning made you late by five minutes and put you in the middle of a traffic accident that sets in motion dozens of other events that have consequences with further implications for years to come.

Think of all your actions, however small. They all have reactions and consequences that get amplified over time. Think of the job you could be taking on because your bid was the lowest. It could end up being a catastrophe while the owner's home burns to the ground leaving your business bankrupt. Due to that, you have to go take a low paying job and who knows where your life goes, never mind the homeowner's life and their family. What if you fire your carpenter tomorrow and his kids can't go to college, as he never finds another job; his children's children then never go to college and generations later God knows what

193

has happened. What if you were ten seconds late to the hardware store and that beautiful girl had already walked out? You would have never met your wife, never had those kids, or grandkids. Maybe you would have had others. The statistical implications of all our minute actions are mind-boggling. Think about your daily dealings and their impact on the world and the future. To quote one of my favorite movies, *Gladiator*, "What we do in life echoes in eternity." Your very small deed can have perpetual life changing implications for generations to come. So walk through this life knowing that you are not just a muted pulse in a universe of noise, but that your every day makes a difference for days on end.

With mathematics on my mind, I have to end the book by giving proper credit where it is due. Being the most humble and good-hearted people I have yet to know, both my mother and father are mathematicians. I owe them my quick thinking and logical mind. In fact, I owe them much more. I would be nothing without my loving parents, literally and metaphorically. Through me, they have passed on all the good they have in themselves, to which I attribute all my relentless spirit, drive, and success. I am forever grateful to Ion and Viorica Fatu.

I am also very appreciative of your patience and consideration in reading my book. I have set a bar that I hope you will challenge yourself to reset higher. Thank you.

Cheers to your success!

INDEX

Books from Allworth Press

Allworth Press is an imprint of Skyhorse Publishing, Inc. Selected titles are listed below.

Brand Thinking and Other Noble Pursuits
by Debbie Millman (6 x 9, 320 pages, paperback, $19.95)

Business and Legal Forms for Graphic Designers
by Tad Crawford and Eva Doman Bruck (8 ½ x 11, 256 pages, paperback, $29.95)

Corporate Creativity: Developing an Innovative Organization
by Thomas Lockwood and Thomas Walton (6 x 9, 256 pages, paperback, $24.95)

Effective Leadership for Nonprofit Organizations
by Thomas Wolf (6 x 9, 192 pages, paperback, $16.95)

From Idea to Exit: The Entrepreneurial Journey
by Jeffrey Weber (6 x 9, 272 pages, paperback, $19.95)

Infectious: How to Connect Deeply and Unleash the Energetic Leader Within
by Achim Nowak (6 x 9, 256 pages, paperback, $19.95)

Intentional Leadership
by Jane A. G. Kise (7 x 10. 200 pages, paperback, $19.95)

Peak Business Performance Under Pressure
by Bill Driscoll (6 x 9, 224 pages, paperback, $19.95)

The Pocket Small Business Owner's Guide to Building Your Business
by Kevin Devine (5 ¼ x 8 ¼, 256 pages, paperback, $14.95)

The Pocket Small Business Owner's Guide to Business Plans
by Brian Hill and Dee Power (5 ½ x 8 ¼, 224 pages, paperback, $14.95)

The Pocket Small Business Owner's Guide to Negotiating
by Kevin Devine (5 ½ x 8 ¼, 224 pages, paperback, $14.95)

The Pocket Small Business Owner's Guide to Starting Your Business on a Shoestring
by Carol Tice (5 ½ x 8 ¼, 244 pages, paperback, $14.95)

The Pocket Small Business Owner's Guide to Taxes
by Brian Germer (5 ½ x 8 ¼, 240 pages, paperback, $14.95)

To see our complete catalog or to order online, please visit *www.allworth.com*.